AF531432

BACTERIAL PLANT DISEASES

By

Dr. Shubhrata R. Mishra

Department of Botany
Vikram University
Ujjain (M.P)

DISCOVERY PUBLISHING HOUSE
NEW DELHI

First Published - 2003

Reprinted - 2016

ISBN: 978-81-7141-749-0

Bacterial Plant Diseases

Published by:

DISCOVERY PUBLISHING HOUSE PVT. LTD.

4383/4B, Ansari Road, Darya Ganj

New Delhi-110 002 (India)

Phone: +91-11-23279245, 43596064-65

Fax: +91-11-23253475

E-mail: discoverypublishinghouse@gmail.com

sales@discoverypublishinggroup.com

web: www.discoverypublishinggroup.com

Printed at:

Infinity Imaging Systems

Delhi

Preface

I have tried to present the significant information about bacteria and their plant diseases in a readable and intelligible way for both under-graduate and post-graduate students. Bacteria are a completely separate group of living organisms, which can live in the most extra-ordinary places. This book is designed so as to acquaint the student with the history, occurrence, structure, special activities like gram reaction, reproduction and genetic recombination, locomotion, endospore formation etc., special DNA characteristics like nucleoids, plasmids etc., classification and some important bacterial plant diseases of bacteria. The book, in its present form, covers the syllabi of most of the Indian Universities.

This book is not an outcome of author's original research but is a compilation work incorporating the researches of the bacteriologists in India and abroad. The matter has been compiled from standard texts, reviews, research papers and websites; most of the figures have been adapted from books and websites.

I would like to express my gratitude to the authors and the publishers of standard texts, research papers and reviews from where the matter of this book has been compiled.

I wish to extend my humble thanks to Professor M.P. Arora for his valuable help and encouragement. My thanks are due to my friends, colleagues for their support and helpful suggestions.

My greatest debts are to my mother who always taught me to complete the every task with honesty, sincerity and dedication and to my husband Dr. Ravi Mishra for his understanding and continuous encouragement without which the book would not have been written.

Last but not the least I offer my thanks to Mr. Indresh Gupta for typing of this book.

Despite my sincere efforts it is possible that some important information might have escaped my attention. Also, students and teachers might feel the need for further details about bacteria. I shall be obliged if the readers would send me their valuable suggestions in this regard or any other aspect of the book. These would help me to improve the book.

Dr. Shubhrata R. Mishra

Contents

Preface

1. **Introduction** 1
History
Shape
Flagella
Methods of Respiration
Mode of Nutrition
Gram Positive and Gram Negative Bacteria
Bacterial Reproduction and Genetic Recombination
Impact of External Environment on Bacteria
The Atmospheric and Temperature Requirements of Bacteria

2. **The Structure of Bacteria** 13
Introduction
Cytoplasmic Membrane
Bacterial Cell Wall
Bacterial Cytoplasm
Mesosomes

3. **Classification of Bacteria** 27
Introduction
Nineteen Sections of Bacterial Group by Bergey's Manual
Groups of Bacteria
Bacterial Nomenclature

4. **Special DNA Characteristics of Bacteria** 58

Introduction

Large Scale Isolation of Plasmid and Bacterial DNA

Plasmids

R-Plasmids or R-factor

F-factor or Sex Factor

Control of Gene Expression in Bacteria

Gene Discovery in Domain Bacteria

5. **Special Activities of Bacteria** 79

Introduction

The Gram Reaction

The Modes of Motion

The Spore Formation

Genetic Recombination

6. **Bacterial Diseases in Plants** 145

Introduction

Common Symptoms of Bacterial Plant Diseases

Common Control Measures

References 197

Index 211

Chapter—1 Introduction

History

Bacteria, often just called 'microb' (micro meaning very tiny) are incredibly small and incredibly common. A powerful microscope that magnifies about 10,000 times is used to see the bacteria. Bacteria are the simplest life forms which are microscopic and show the characteristics of both plants as well as animals. They were discovered by Antony Von Leeuewenhock in 1676. In 1675, he was first time made notable observations on bacteria using crude microscopes that could magnify 150 times. The importance of these observations was, however, realized only after Pasteur (1876) had demonstrated their role in fermentation and decay and Koch (1876) had proved that bacteria can cause such diseases as anthrax, tuberculosis or asiatic cholera. The name bacteria was given by Ehrenberg.

Bacterial Abundance

High Diversity

- A single gram of rich, undisturbed soil may contain as many as 5,000 different species of bacteria;
- In fact, the health, and therefore the biological diversity of a given environment may be measured, in part, in terms of the sheer numbers of bacterial species present, per gram.

High Total Numbers

- The total number of bacteria of all types present in that gram of soil numbers in the billions (10^9);
- Though this number is huge, the concentration of bacteria in mammalian feces is hundereds of times even greater;

- The total number of bacteria living in or on the average human consist of more cells than the total number of human cells making up the human body;
- Bacteria numbers, in short, are immense.

It is now well recognized that of all the organisms, bacteria are most closely related to man's life. T.J. Burrill of the University of Illinois (1878) was the first to prove the association of a bacterium with a plant disease. There are about 1600 species of bacteria of which 50% are true bacteria. Excepting in pits of volcanoes and deep strata of rocks, bacteria are present almost everywhere. Just one teaspoon of water from a river or the sea normally would contain 50 million bacteria. Bacteria used to be thought to be primitive plants. Now scientists know that they are a completely separate group of living things which can survive in the most extra-ordinary places like extrernes of temperature, pH, oxygen tension, and osmotic and atmospheric pressures. Some can even live in boiling water; others live inside us animals and help us digest food. They were probably the first life on Earth.

Shape

Bacteria are various shapes. They may show more than one shape. Thus they exhibit pleomorphism Bacteria occur in three main shapes:

1. **Spherical Bacteria:** They are called cocci (singular coccus). The cells may occur in pairs (diplococi), in groups of four (tetracocci), in bunches (Staphylococci), in a bread- like chain (streptococci) or in a cubical arrangement of eight or more (sarcinae).
2. **Rod-like Bacteria:** They are called bacilli (singular bacillus). They generally occur singly, but may occasionally be found in pairs (diplobacilli) or chains (strepto bacilli).
3. **Spiral-shaped Bacteria:** They are called spirilla (singular, spirillum). Short incomplete spirals are called vibrios or comma bacteria.

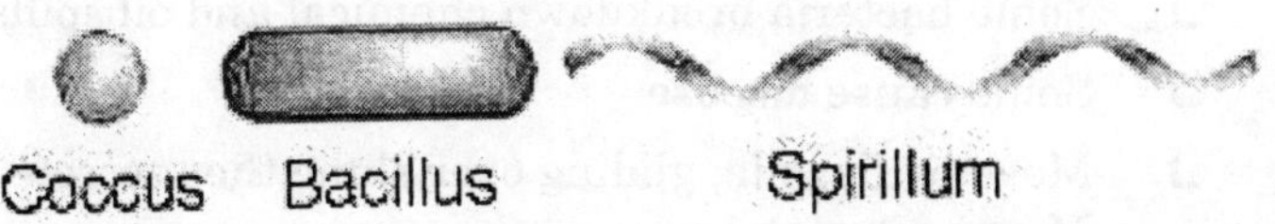

Fig. 1: Types of Bacteria

Although bacteria are of three basic shapes, they display an astonishing variety of forms when viewed microscopically. Some irregular shapes are also found in bacterial cells. For example- *Rynebacteria*, L-phase of *Agrobacterium* and *Erwinia* have V-Form. The L-Form or L-phase of these bacteria lack rigid cell wall. The shape of bacterial cell is of fundamental importance in the classification and identification of bacteria.

Bacterial Evolution

- Most numerous organisms on earth
- Earliest life forms (Fossils date 2.5 billion years old)
- Microscopic prokaryotes (no nucleus nor membrane-bound organelles)
- Contain ribosomes
- Infoldings of the cell membrane carry on photosynthesis and respiration
- Surrounded by protective cell wall containing peptidoglycan (protein-carbohydrate)
- Many are surrounded by a sticky, protective coating of sugars called the capsule or glycocalyx (can attach to other bacteria or host)
- Have only one circular chromosome
- Have small rings of DNA called plasmids
- May have short, hairlike projections called pili on cell wall to attach to host or another bacteria when transferring genetic material
- Most are unicellular
- Found in most habitats
- Most bacteria grow best at a pH of 6.5 to 7.0
- Main decomposers of dead organisms so recycle nutrients

- ❑ Some bacteria breakdown chemical and oil spills
- ❑ Some cause disease
- ❑ Move by flagella, gliding over slime they secrete (e.g. Myxobacteria)
- ❑ Some can form protective endospores around the DNA when conditions become unfavourable; may stay inactive several years and then re-activate when conditions favourable.

Flagella

Hair like projections are present in bacteria which is known as flagella. When the flagella are present, the bacteria are motile otherwise they are non-motile. The flagella are delicate and they show variation in length and texture. They are unique structures having range in thickness from 15 nm to more than 24 nm, and consist of single or entwined structures that resemble microtubules but are composed of a single protein called flagellin. A structure some what analogous to the flagellum but which is not involved in motility is the pilus (plural, pili) or fimbrae. Pili are considerabely shorter than flagella and are more numerous on the cell. Certain pilli are involved in the transfer of genetic material from one bacterial cell to another.

The flagella are borne in a specific manner and their number is usually constant for a species. On the basis of number and arrangement of flagella, there are following types of bacteria:

1. **Atrichous:** Bacteria without flagella.
2. **Monotrichous:** Bacteria having only one flagellum at one end of the cell.
3. **Lophotrichous:** Bacteria having two or bunch of flagella at one pole of the cell.
4. **Amphitrichous:** Bacteria having bunch of flagella at both the poles of the cell.
5. **Peritrichous:** Bacteria having flegella all over the cell surface.

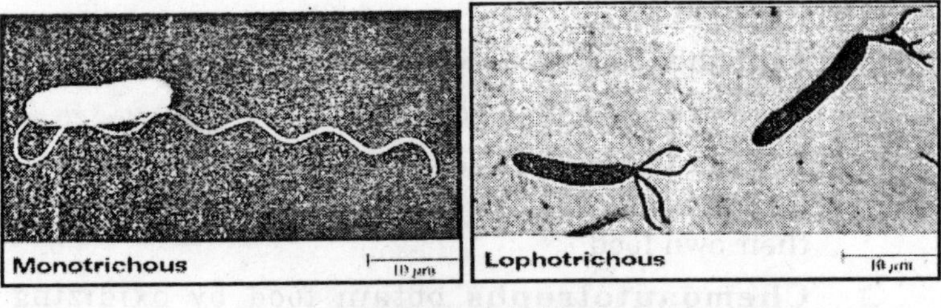

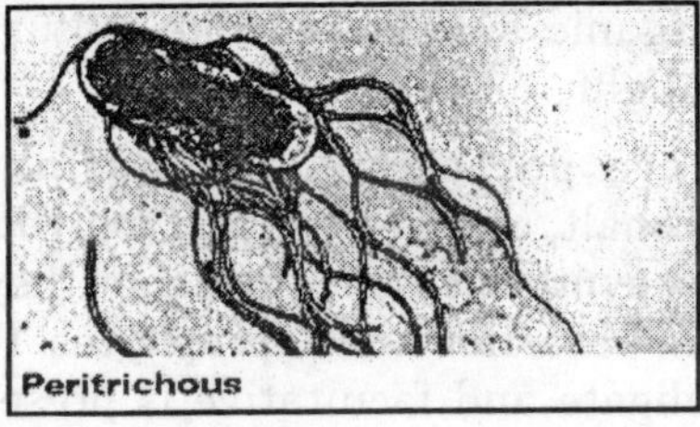

Fig. 2: Types of flagellation in Bacteria

Methods of Respiration

Both types of respiration methods e.g., aerobic and anaerobic respiration are found in bacteria. Obligate aerobic bacteria can't live without oxygen; (Tuberculosis Bacteria). Obligate anaerobes die if oxygen is present; (Tetanus bacteria that causes lackjaw). Facultative anaerobes do not need oxygen, but don't die if oxygen is present; (*E.coli*). Anaerobes carry on fragmentation, while aerobes carry on cellular respiration.

Mode of Nutrition

Like any other living organisms, bacteria also require nutrients from the environment for the synthesis of their cell material and for the generation of energy required for running the machinary for bio-synthesis. The chemical composition of cells, broadly constant thought the living world, indicates their major material requirement. The major elements required in cell nutrition are hydrogen, oxygen, carbon, nitrogen, phosphorus and sulphur, in order of decreasing abundance in the cell. These six elements account for 95 per cent of the cell dry weight. Many other elements are required in traces, in the remaining 5 per cent, but are essential.

Methods of Nutrition

- **Saprobes** feed on dead organic matter
- **Parasites** feed on a host cell
- **Photoautotrophs** use sunlight for energy, but get carbon from organic compounds (not CO_2) to make their own food
- **Chemoautotrophs** obtain food by oxidizing inorganic substances like sulfur, instead of using sunlight

These are potassium, magnesium, calcium, iron, manganese, cobalt, copper, zinc and molybdenum. Most of the bacteria do not synthesize their food thus they are heterotrophic. They are both, parasites and saprophytes. Parasititism of both kinds, i.e. obligate and facultative is present in the bacteria. Like wise the bacteria are both obligate and facultative saprophytes. Some bacteria are very specific also in their food requirements. Thus they can utilize only a few specific organic compounds. Bacteria also synthesize enzymes, thus they digest their food themselves. The nutritional classification, based on energy source and principal carbon source, divides bacteria into following four major groups:

1. **Photosynthetic Bacteria:** Using sun light as the energy source and carbon dioxide as the principal carbon source. The photosynthetic process in bacteria differs from the photosynthesis of higher green plants. The photosynthetic bacteria possess chlorophylls (bacteriochlorophylls) different from those of plants in structure as well as in light absorbing properties. Bacteriochlorophylls absorb light in the near infrared region (660-870 nm). They are not contained in chloroplasts but are found in extensive membrane systems. Oxygen is not liberated during the photosynthetic process and the hydrogen source is not water but usually sulphur compound (e.g. hydrogen sulphide). Example: Purpul non sulphur bacteria, *Cloribium limicola*.
2. **Photochemosynthetic Bacteria:** Using light as the energy source and an organic compound as the

principal carbon source. Example: Purpul and Green Bacteria.

3. **Chemoautotrophic Bacteria:** Using a chemical energy source and carbon dioxide as the principal carbon source, energy is obtained by oxidation of reduced organic compounds. Only certain bacterial types belong to this group.

4. **Chemoheterotrophic Bacteria:** Using a chemical energy source and an organic compound as the principal carbon source. Most bacteria, all the pathogenic types, belong to this group. Both energy and carbon can usually be derived from the metabolism of a single compound.

Plant pathogenic bacteria are basically facultative saprophytes, i.e., in the absence of the living host plant tissue, they can remain active as saprophytes on dead organic matter also. Thus, most of the true bacteria can be grown on artificial culture media, the composition of which may have to be varied according to some specific requirement of the bacterium. The degree of saprophytic survival ability varies widely depending on the resistance of the bacterium to biological competition and deleterious biochemical agents in soil.

Gram Positive and Gram Negative Bacteria

The bacteria are of two kinds on the basis of their wall structure and their response to a differential stain prepared by the Danish bacteriologist Hans Gram (1884). The stain is a

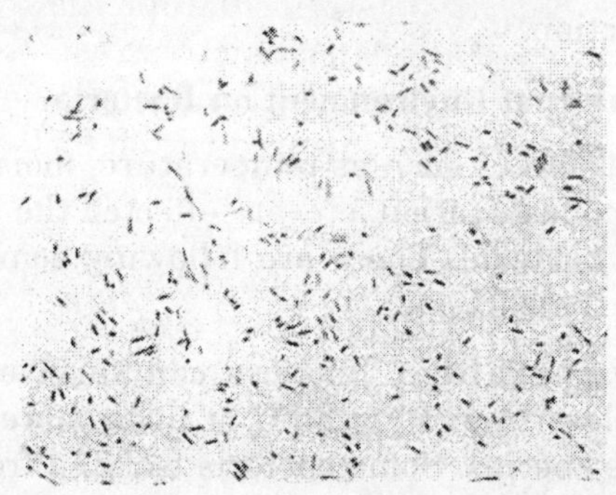

Fig. 3: Gram stain of Gram–*E. coli* cells

combination of crystal violet solution and an iodine solution. The crystal violet dye stains the bacteria blue. When these are subsequently treated with iodine and alcohol the bacteria which retains the stain are called Gram-Positive while others which do not retain the stain are called Gram Negative. This technique is used to identify bacteria.

Fig. 4: A Gram stain of Gram+*Staphylococcus* cells.

Bacterial Reproduction and Genetic Recombination

Most bacteria reproduce asexually by binary fission in which chromosome replicates and then the cell divides. Bacterial replicate (double in number) every 20 minutes under ideal conditions. Bacteria contains much less DNA than eukaryotes. Bacterial plasmids are used in genetic engineering to carry new genes in to other organisms. Bacteria recombine genetic material in three ways—transformation, conjugation and transduction.

Impact of External Environment on Bacteria

Physical forces such as temperature, moisture pressure, light, H-ion concentration, etc. are affected the vital metabolic processes of bacteria. There are following some major forces which affect the bacteria:

1. **Temperature:** Bacteria can survive temperatures of less than 0° to 85°C or even more depending on the species. Some bacteria isolated from hot springs are capable of growth at temperatures as high as

95°C, others isolated from cold environments can grow at temperatures as low as –10°C, if the medium does not freeze.

For every organism there is a maximum, a minimum and an optimum temperature range for growth. Growth temperature range or cardinal temperature is the range of temperature between the minimum and the maximum up to which bacterial activity will occur. Above and below the optimum, the bacterial growth declines and ultimately ceases at the extremes, on the basis of growth temperature range bacteria can be broadly grouped in following three categories:

Psychrophilic Bacteria: Bacteria which can active at the minimum temperature 0°C, optimum temperature 15°C and maximum temperature 30°C.

Mesophilic Bacteria: Bacteria which can survive at the minimum temperature range 5-25°C, optimum temperature 20-45°C and maximum 30-50°C.

Thermophilic Bacteria: Bacteria which can survive at the minimum temperature 15-45°C, optimum temperature 55°C and maximum temperature 60-93°C.

Thermal death point or thermal death rate is the temperature at which all bacterial cells in a solution after an exposure of 10 minutes are killed. Similarly, thermal death time is the time required to kill all cells of a bacterial species at a given temperature.

Only vegetative cells are considered for these determinations. Spore forming bacteria survive even the boiling temperature. To kill them the spores must be activated to grow into vegetative cells. Such bacteria have two thermal death points or time, one for vegetative cells and one for spores. Coagulation of proteins in the bacterial protoplasm causes the death of bacteria at highest temperature. This coagulation is governed by many factors such as water content of the medium, water content of the organism, pH of the medium, composition of the medium, age of cells, etc. The greater the percentage of water in a medium, the lower will be the temperature required to kill bacteria. Moist heat is a more effective agent of sterilization than dry heat. In autoclaves, moist heat at comparatively lower temperature gives perfect sterilization within 15 to 20 minutes than dry heat in an over where time required for sterilization may be in hours. Dry bacterial cells are more tolerant to heat than young actively growing cells with more moisture. Killing of bacteria is rapid if the reaction of the medium is alkaline or acidic. Media containing high quantities of protein or albuminous materials require high temperatures for death of bacterial cells suspended in them. The protein, which forms a protective film around the cells, thus hinders penetration of heat.

2. **Moisture:** Bacteria are comparatively more aquatic than terrestrial and survive better in the presence of a high percentage of water. However, in water-logged soils where availability of free oxygen is reduced the strictly aerobic bacteria cease to multiply and their population is replaced by anaerobic bacteria. If bacterial cells are dried they may cease activity but may survive and when restored to moist conditions regain their activity. Water mainly helps bacterial cells by facilitating diffusion of nutrients through cell membrane and by providing a fluid medium for movement. However, survival of plant pathogenic bacteria is generally reduced if the soil is wet. It is because in wet soil microbial activity is increased and this encourges antagonism.

3. **Light:** Bacterial activity is not adversely affected by ordinary visible light. Ultraviolet rays and infra red rays of light are bactericidal. Strongly bactericidal rays are those between 2,500 and 2,800Å.

4. **Pressure:** Bacterial cells are not so sensitive to ordinary mechanical pressure due to their minute size and volume. Osmotic pressure is the best known pressure on which bacteria can survive. The fluid inside the bacterial cell is always hypertonic in relation to the fluid out side in the host cell or in culture medium. Due to osmotic pressure the fluid from out side enters the cell. This is how nutrients

enter the bacterial cell. When the bacterial cells are placed in a hypotonic solution, water enters the cell due to plasmolysis, so the cell is swelled. Ultimately, the cell-wall may burst.

The Atmospheric and Temperature Requirements of Bacteria

Some bacteria have an absolute requirement for oxygen. These are the obligate aerobes. Others, the facultative anaerobes can survive in the absence as well as the presence of oxygen. The obligate anaerobes are killed by traces of oxygen. A small group of bacteria are killed by normal atmospheric levels of oxygen, but yet require traces of oxygen to grow. These are the referred to as microaerophiles.

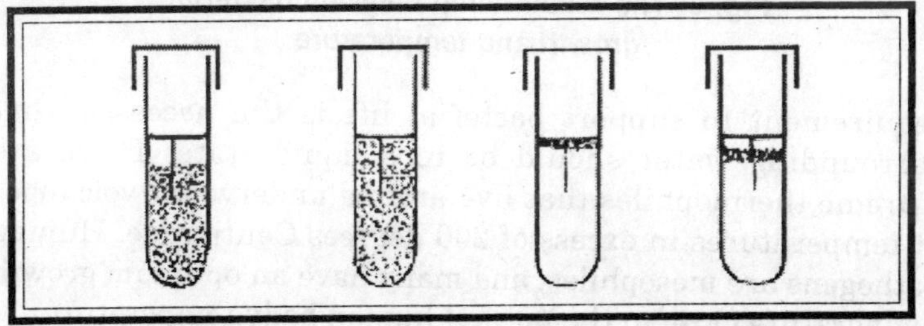

Fig. 5: The relationship between bacterial growth and oxygen

The culture on the left is an obligate anarobe, unable to grow in the presence of oxygen, close to the surface of the growth medium. Next is a facultative bacterium, which is indifferent to the presence of oxygen. There then follows an obligate aerobe that can only grow at the surface of the culture medium. The culture on the right is a microaerophile. It can only grow where the oxygen tension is low.

Bacteria that grow at very low temperatures are known as psychrotrophs. Bacteria found to grow at high temperatures are known as thermophiles. Those that grow at moderate temperatures are known as mesophiles.

It is not possible to determine the minimum temperature at which some psychrophiles can grow. The laboratory media used to determine these temperatures becomes frozen during the experiments. It would seem probable that the only

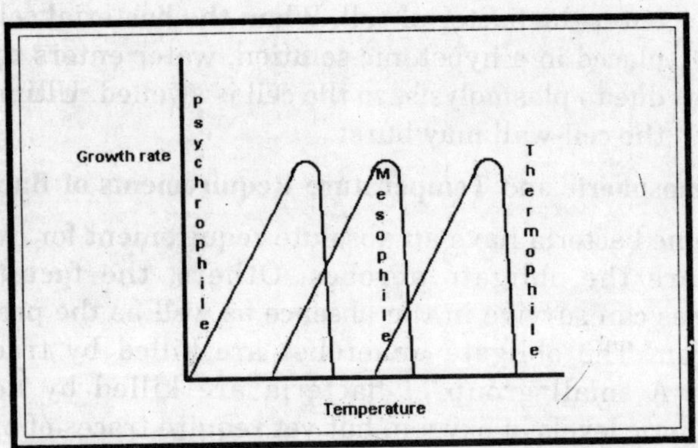

Fig. 6: The relationship between bacterial growth and temperature

requirement to support bacterial life is the necessity that surrounding water should be in a liquid state. There are extreme thermophiles that live around underwater volcanoes at temperatures in excess of 200 degrees Centigrade. Human pathogens are mesophiles, and many have an optimum growth temperature close to the normal human body temperature.

Chapter—2 | The Structure of Bacteria

Introduction

Bacteria are pokaryotic in their organization. The bacterial cell is surrounded by a capsule which encloses the nuclear structure or incipient nucleus. They do not have membrane-bounded organelles or a well organised nucleus and their cell wall composition is entirely different. Beneath the capsule is a cell wall which gives shape to the cell. Cell wall is very rigid and plays a vital role in bacterial biology. It consists of lipids, carbohydrates, proteins, phosphorus and other inorganic substances. Here the rigidity is due to a unique polymer, the mucopeptide, which is based upon a backbone of alternating molecules of N-acetyl glucosamine and N-acetyl-muramic acid. Attached to the muramic acid are short chain molecules of aminoacids which are cross linked to other muramic acid molecules in adjacent chains. This cross linking enables the complex to form a rigid network. At times the slime of cell wall becomes thick, gummy and mucilaginous this is known as capsule. The capsule is a resistant covering against the unfavourable conditions. Capsules and slime layers of certain bacteria are made up of polysoccharides, a material composed of dextrans, levans, etc. and appear like plant gums. Capsules or slime of certain other bacteria are composed of polypeptides and peptides.

The cytoplasm is granular, viscous and lies between the nuclear material and plasma membrane. The protoplasm of bacterial cell is similar to that of other living organisms. The moisture content is 70-85%, and is colloidal in nature. The cytoplasm is densely packed with ribosomes of 70S. The definite components of nucleus like chromonemata, nucleolus and nuclear membrane are not found in the bacterial nucleus.

The difficulty of demonstrating the nuclear material in bacterial cells is due to the extreme basophily of the cells because of the presence of RNA in the cytoplasm. Most of the bacteriologists now believe that the nucleus in bacteria is present in the cell as a discrete body. The chromatinic structures of bacteria can be distinguished from the cytoplasm of the cell. These structures increase in size and usually give rise to short rods which multiply by splitting in a lengthwise manner. More recently the bacteriologists have demonstrated the staining of a body which takes the stain of DNA.

Structure	Function
Cell Wall	Protects the cell and gives shape
Outer Membrane	Protects the cell against some antibiotics (only present in Gram Negative Cells)
Cell Membrane	Regulates movement of materials into and out of the cell; contains enzymes important to cellular respiration
Cytoplasm	Contains DNA, ribosomes, and organic compounds required to carry out life processes
Chromosome	Carries genetic information inherited from past generations
Plasmid	Contains some genes obtain through genetic recombination
Capsule, and slime layer	Protects the cell and assist in attaching the cell to other surfaces
Endospore	Protects the cell against harsh environmental conditions, such as heat or drought
Pilus (Pili)	Assist the cell in attaching to other surfaces, which is important for genetic recombination
Flagellum	Moves the cell

Bacterial cell has a single chromosome, which consists of two chains of DNA wound round each other. The chromosome is ring shaped and it is embedded in the cytoplasm. This whole body usually rests in the centre of the cell. The total length of the DNA strand is proportional to its molecular weight and is often much longer than the length of the cell.

Other intracellular organelles like golgibodies, nucleolus, endoplasmic reticulum, lysosomes and mitochondria are absent in bacteria. Microtubules have been reported in L-form of certain bacteria. On the other hand, extrachromosomal genetic elements such as plasmids and episomes are characteristic organelles in many bacteria.

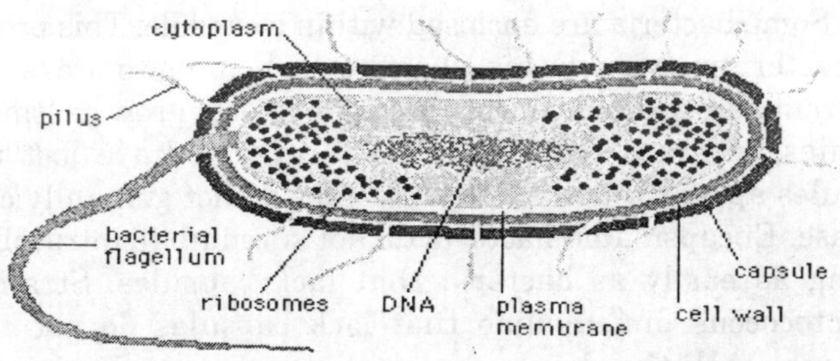

Fig. 7: Generalized diagram of a bacterium

Capsules: The capsule is not essential for the survival of bacteria under favourable growth conditions. However, it protects the bacteria against adverse conditions. The capsule may also function in the storage of food substances, and as a storage site for the disposal of waste material. Capsules may be divided into two categories:

Macrocapsules: They are at least 0.2 μm thick and can be seen under the light microscope.

Microcapsules: They can not be seen under the light microscope, but can be demonstrated immunologically.

The capsule is secreted by the bacterial cell and is usually composed of polysaccharides or disaccharides, but is sometimes made up of polypeptides. Most bacterial polysaccharides contain more than one type of sugar, i.e. they are heteropolysaccharide, Bacterial capsules are species—specific, and can therefore be used for immunological distinction of closely related species. The daxtrans produced by *Leuconostoc mesenteroides* are used as plasma expanders in hospitals. They are also used as laboratory gels for filtration. In *Acetobacter xylinum* the polysaccharide is a homopolysaccharide made up of glucose units linked β1-4. *Pseudomonas aeruginosa* secretes a heteropolysaccharide consists of D-glucose, D-galactose, D-mannose, D-glucuronic acid and L-rhamnose residues. The capsule of pneumococci, is made up of hexoses, uronic acids and amino sugars. The protein capsule consists of L-amino acids in the streptococci. Polypeptides in the capsule of *Bacillus anthracis* contain D-glutamic acid.

Some bacteria are enclosed within a capsule. This protects the bacterium, even within phagocytes, helping to prevent the cell from being killed. Encapsulated bacteria grow as 'smooth' colonies, whereas colonies of bacteria that have lost their capsules appear rough. Rough colonies do not generally cause disease. Encapsulated bacteria do not succumb to intracellular killing as easily as bacteria that lack capsules. Strains of Streptococcus pnuemoniae that lack capsules do not cause disease. All the bacteria that cause meningitis are encapsulated. Suspending bacteria in India ink is an easy way of demonstrating capsules. Ink particles cannot penetrate the capsular material and encapsulated cells appear to have a halo around them. This is the Quellung reaction.

In the 'Quellung' reaction, bacterial cells are resuspended in antiserum that carries antibodies raised against the capsule. This causes the capsule to swell, and this can be easily visualised by suspension in India Ink. The ink particles cannot penetrate the capsule, which this appears as a halo around the bacterial cells.

Some bacteria produce slime to help them to stick to surfaces. Slime is produced by several types of pathogenic microbes, and is usually made up from polysaccharides. The slime produced by Streptococcus mutans enables it to stick to the surface of teeth, where it helps to form plaque, leading

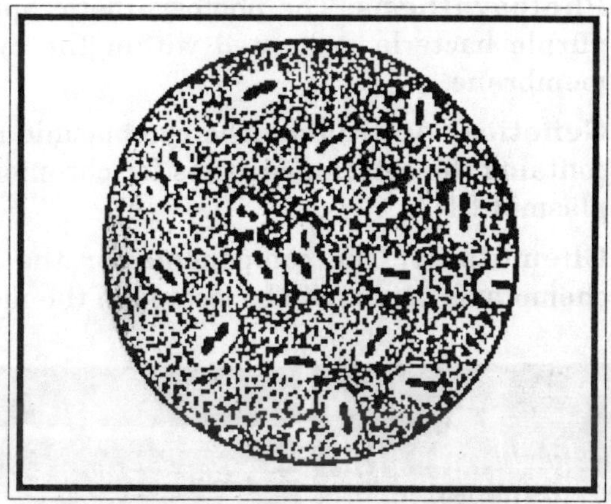

Fig. 8: The Quellung reaction

eventually to dental caries. 'Coagulase-negative' staphylococci live on the skin, and some strains produce a slime that enables them to stick to plastics. These bacteria cause infections associated with implanted plastic medical devices.

Cytoplasmic Membrane

In gram negative bacteria beneath the cell wall an inner membrane is present. This is called cytoplasmic membrane. Cell wall and cytoplasmic membrane are separated by 100Å space called the periplasmic region. This space contains a peptidoglycan layer. The cytoplasmic membrane of bacterial cell is an important centre of following metabolic activities:

1. **Synthesis of Cell Wall Components:** The membrane has the enzymes of the biosynthetic pathways which synthesise the components of cell wall such as phospholipids, peptidogly cans, teichoic acids, lipopolysaccharides and simple polysaccharides.
2. **Transport of Nutrients:** The permeases are responsible for the transport of organic and inorganic nutrients through the membrane.
3. **Electron Transport and Oxydative Phosphorylation:** The membrane of aerobic bacteria contains the components of the electron transport chain and oxidative phosphorylation.

4. **Photosynthesis:** The photosynthetic apparatus of purple bacteria is located within the cytoplasmic membrane.
5. **Genetical Approach:** The cytoplasmic membrane contains the attachment sites for chromosomal and plasmid DNA.
6. **Chemotaxis:** The components for the control of chemotaxis appears to be located in the membrane.

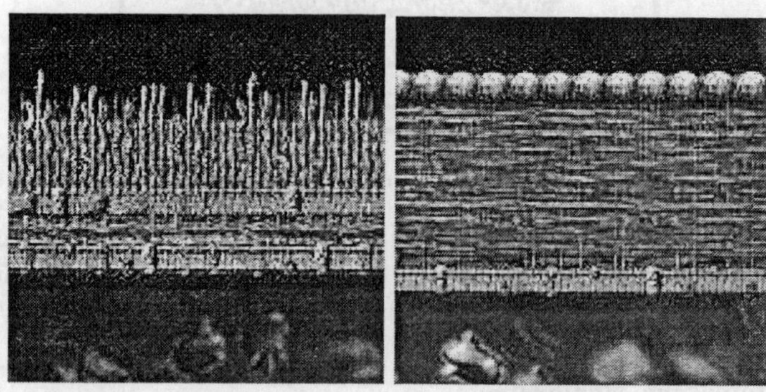

Gram Negative Gram Positive

Fig. 9: Cell wall and cytoplasmic membrane of Gram Negative and Gram Positive Bacteria

The major characteristics of the structure of cytoplasmic membrane is explained by the Fluid-mosaic model of Singer and Nicolson (1972). According to this the biological membranes are regarded as the fluid mosaic or quasifluid structures. In such membranes the lipids and integral proteins, lipid-protein and lipid-lipid interactions contributed to membrane structures and dynamics. The proteins of the membrane are concerned with the enzymatic activity of the membrane and also with the transport of molecules across the membrane. As for the lipid layer it provides the permeability barrier. This model also envisages two broad categories of proteins, peripheral and integral types. Integral proteins are supposed to be very active where as peripheral proteins are located superficially at the membrane surface.

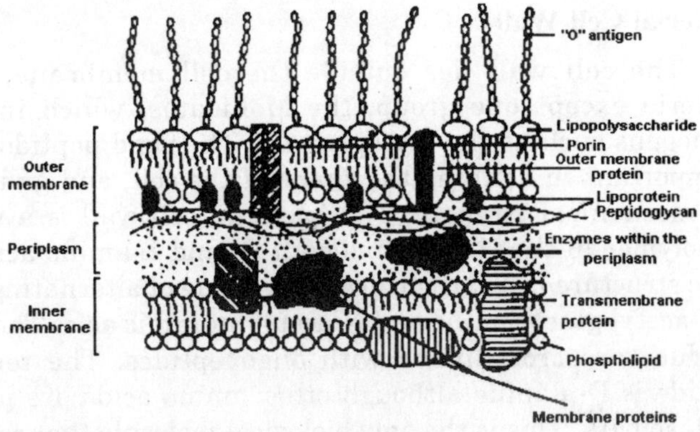

Fig.10: Fluid-mosaic model (Nicolson and Singer)

In the Fluid-mosaic model the phospholipid bilayer is a cementing framework with attached proteins. Some of the proteins are embedded within the bilayer.They visualized two types of covalent interactions to be predominant. These are hydrophobic and hydrophilic. Both the interactions are maximized. Interactions existing among the charged amphipathic proteins, the water phase and the polar heads of amphipathic phospholipids. Hydrophobic iteractions exist between uncharged protein and nonpolar tails of phospholipid bilayer. Because of the rapid movement of the lipid and protein molecules, the Singer-Nicolson membrane is considered to be highly fluid.

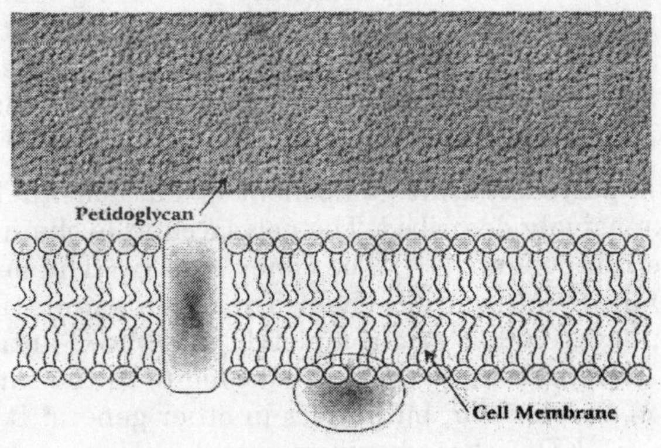

Fig. 11: The Gram positive cellwall

Bacterial Cell Wall

The cell wall lies outside the cell membrane, in all bacteria except one group, the Mollicutes, which includes pathogens such as the mycoplasmas. The rigid peptidoglycan is important in defining the shape of the cell, and giving the cell mechanical strength. The bacterial cell wall is a unique biopolymer in that it contains both D- and L-amino acids. Its basic structure is a carbohydrate backbone of alternating units of N-acetyl glucosamine and N-acetyl muramic acid. The NAM residues are cross-linked with oligopeptides. The terminal peptide is D-alanine although other amino acids are present as D-isomers. This is the only biological molecule that contains D-amino acids and it is the target of numerous antibacterial antibiotics. The cell wall of Gram-positive bacteria lies beyond the cell membrane and is largely made up of pepidoglycan. There may be up to 40 layers of this polymer, conferring enormous mechanical strength on the cell wall. Other polymers including teichoic and teichuronic acids also lie in the cell walls of Gram-positive bacteria. These act as surface antigens. Cell wall is morphologically similar to the cytoplasmic membrane but contains less phospholipid and fewer proteins.

1. **Carbohydrate Components:** It contains a unique carbohydrate component called lipopoly- saccharide (LPS). LPS consist of complex polysaccharides covalently linked to lipid A. Lipid A contains glucosamine, phosphate and fatty acids. The carbohydrate component of lipid A consists of β-1, 6-D-glucosamine disaccharide units. Polysaccharide has antigenic properties while lipid A has endotoxin properties.

The polysaccharide component of *Salmonella* LPS has been extensively described. The polysaccharide chain consists of three components - (i) The inner core (ii) The outer core and (iii) the O-antigen side chain. The rough strains (R Forms) of gram negative bacteria completely lack the O-antigen side chains. The inner core region is believed to be the same in all strains of *Salmonella*, but differs in other genera. It has two regions (i) KDO (ketodeoxyoctonate) region and (ii) the diheptose region. The KDO region consists of three units of

KDO. The diheptose region consists of two units of Hep (L-glycero-D-mannoheptose sugar) KDO and Hep are unique to bacterial cell walls.

2. **Protein Components:** Cell wall has at least four different major protein components; matrix protein, protein 2, tolG proteins and lipoprotein. Matrix protein consists of two distinct peptides, 1a and 1b, which are tightly but non-covalently associated with peptidoglycan. They are also associated with the lipoprotein. Generally many intrinsic proteins have high α-helix contents, but in matrix proteins have high β-structure contents. Low molecular weight nutrients such as sugars, aminoacids and salts probably enter the periplasmic space from the outside through the matrix protein pores. While higher molecular weight nutrients need their own receptors to pass through the membrane. The matrix protein also serve as receptors.

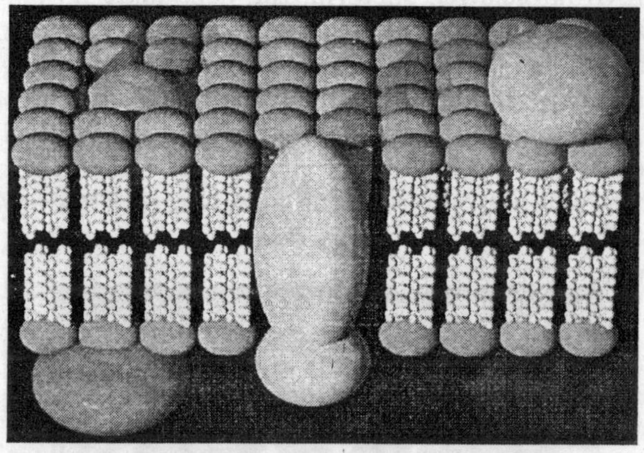

Fig. 12: Lipid bilayer shown in cellwall

Protein-2 is closely related to the matrix proteins. It may be coded by a prophage gene and may function in surface exclusion of super infecting phage. tolG protein is a heat-modifiable and trypsin-sensitive protein. Like the matrix proteins, tolG protein also has a high β structure content. It appears to be exposed to the outer surface of the membrane, and serves as a receptor for certain phages. It has an important function in F-pilus mediated conjugation.

Lipoprotein exists in the free form as well as in the bound form in the cell wall. In E-coli' there are about twice as much bound form molecules as free-form molecules. Lipoprotein has a molecular weight of about 7,000 and consists of 58 amino acid residues. It has a very high α-helical content.

3. **Peptidoglycans:** The vast majority of bacteria have a cellwall containing a special polymer called peptidoglycan. Peptidogycans like mucopeptides, glycopeptides, mureins are the structural elements of almost all bacterial cell walls. Peptidoglycan forms a rigid layer which contributes to the structural integrity of the cell. This is comparatively low in gram negative bacteria. In many gram negative bacteria the peptidoglycan layer is a monolayer, or, at the most, a bilayer.

In Gram positive bacteria, the peptidoglycan makes up 40-90% by dry weight of the cell wall. It is about 30 nm thick in many species. Teichoic acids, teichuronic acids and glycans are three major components found in cell wall of gram positive bacteria. These components are hydrophilic, flexible and linear molecules. Although teichoic acids may show amphoteric properties, but they all are negatively charged. They are linked to peptidoglycan by a single, terminal covalent bond.

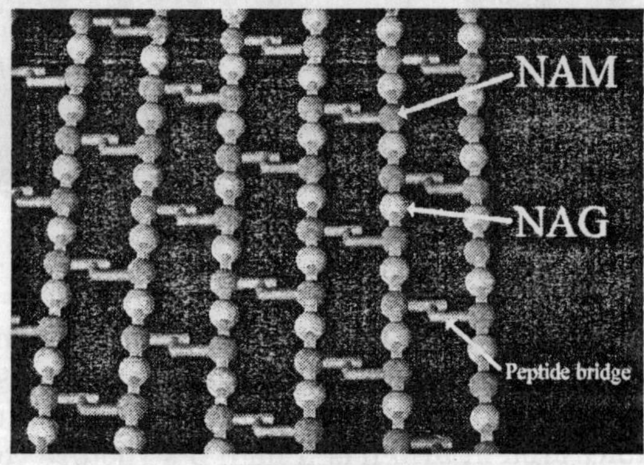

Fig. 13: Peptidoglycans mash

The elastically flexible bacteria, such as the gliding bacteria bend and flex when they hit another object while swimming. However, they return to their original form when the steress is removed. The flexibility is attributed to a low amount of peptidoglycan in the cell wall.

Bacterial Cytoplasm

The cytoplasm of bacteria contains polysomes—a range of ribosomes actively translating messenger RNA in to proteins. Some bacteria also have inclusion bodies within the cytoplasm. Largelly, the cytoplasm of bacterial cell is contained with large granules of storage materials and about 10,000 riibosomes which constitute up to 30% of the weight of the bacterium. The storage granules are high molecular weighted polymers. They serve as stores of energy or organic compounds in the form of readily metabolizable substrates. There are three types of storage granules are found in the cytoplasm i.e., polymetaphosphate granules, poly-β-hydroxybutyrate granules and polyglucan granules. Polymetaphosphate is commonly known as "volutin" granules due to its presence in *Spirillum volutans*. These granules stain reddish violet with methylene blue or toluidine blue, so these are also known as metachromatic granules. Phosphate is stored in these granules in the form of linear chains of inorganic pyrophosphate. These granules are commonly found in *Corynebacteria*, especially *C. diptheriae* and in the *Mycobacteria*.

Poly-β-hydroxy butyrate granules are found in several bacteria. These granules are energy reserve compound. In some bacterial species, e.g., Bacillus megatherium, these granules may make up as much as 60% of the dry weight, especially after growth on acetate or butyrate.

Polyglucan granules are found in certain bacteria like clostridia and coliforms. They stain blue, reddish blue or brown with iodine. These granules consist of glucose units with α 1-4 linkage and occasional α 1-6 branch points.

Ribosomes are found free in the cytoplasm. In functionally active bacteria, the cytoplasm is packed with free and randomly distributed ribosomes. Bacterial ribosome is 70 S ribosome which consists of a larger 50 S sub unit and a smaller 30 S sub unit. 30 S subunits may be·ellipsoid, oblate ellipsoid, triaxial ellipsoid, disc-shaped or obtuse-angled scalene triangle shaped. According to Stofller and Wittmann's model the 30 S subunit has an elongated, slightly bent prolate shape. It is a bipaartite structure. A transverse hollow or cleft divides the 30 S subunit into two parts, a smaller 'head' and a larger 'body', giving it the appearance of a telephone receiver or embryo. In Lake's model the 30 S subunit is considered to be completely asymmetrical. According to this model 30 S sub unit is divided into two unequal parts, the upper one-third (head) and the lower two third (body). Extending from the lower two third is a region called the plateform. There is a cleft between the plateform and the upper one third. This cleft is an important functional region.

50 S subunit also found in many shapes like rounded, kidney-shaped, circular profile with a nose and maple leaf structures. These different shapes have been interpreted as being the same 50 S structure seen in different views. In frontal view, the 50 S subunit appears bilaterlly symmetrical and shows three protuberances arising from a rounded base. Of these the central protuberance is the most prominent. The 50 S subunit has been compared to an armchair, with the rounded base forming a vaulted seat, the central protuberance the back and the lateral protuberance the arms. But according to Lake's model, the 50 S subunit is also asymmetrical.

The 30 S and the 50 S subunits are associated to form the 70 S ribosome. The frontal face of the 30s subunit with its hollow faces the vaulted seat of the 50 S subunit. The long axis of the 30 S subunit is oriented transversely to the central protuberance of the 50 S subunit and the vaulted seat of the large subunit.

Bacterial ribosome for example in *E. Coli* robosome consists of three types of RNA (5 S, 16 S and 23 S) and 53 proteins, The 30 S subunit consists of 16 S RNA and 21 proteins (S 1-S 21) and the 50 S subunit contains 5 S and 23 S RNA and 34 proteins (L1-L34). Two thirds of the mass of the ribosome is RNA. The 5 S RNA of *E. Coli* has 120 nucleotides. It has been observed that 5 S rRNA forms a specific complex with proteins L5, L18 and L25 of the 50 S ribosomal subunit. 5 S rRNA has an important role in binding aminoacyl tRNA to the A-site. 16 S RNA is about 1600 nucleotides long. 16 S RNA can be divided into three specific regions: ***(i) Nucleation site-I***—It is a proximal fragment of about 500 Nucleotides binds to proteins S 4, S 16, S 17 and S 20. ***(ii) Nucleation site-II***—It is a central site of about 550-850 nucleotides interacts with proteins S 6, S 8, S 15 and S 18. ***(iii) Nucleation site -III***—It comprises the 3' one-third of 16 S RNA, and binds to proteins S 6, S 9, S 13, and S 19. 16 S RNA also provides a binding site for tRNA. 235 RNA contains about 3,200 nucleotides. It readily splits into 13 S and 18 S fragments. Individual proteins or groups of proteins can be rebound of these two fragments, enabling localization of the binding sites.

Mesosomes

Mesosomes are complex and localized invaginations of the cytoplasmic (plasm) membrane. These invaginations have many vesicles, tubules or lamellar whorls. These invaginations have many vesicles, tubules or lamellar whorls. Mesosomoes are generally found in gram positive bacteria but rarely also observed in gram negative bacteria. The protein component of mesosome is different that of cell membrane but lipid component is same. Mesosomes were formerly believed to be the equivalents of the mitochondria of higher cells. They were

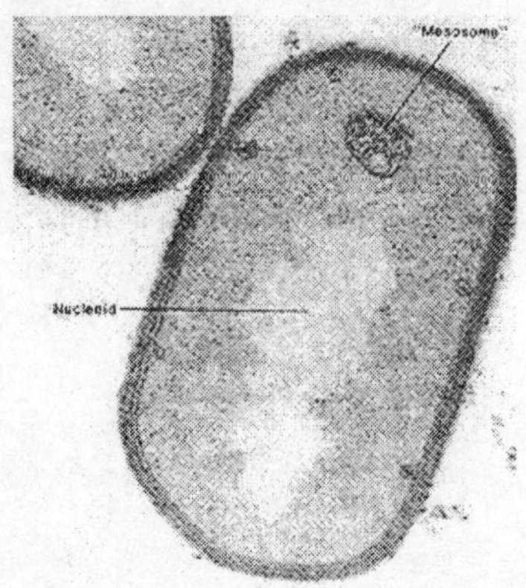

Fig. 14 : An electron micrograph of a mesosome

thought to be the centres of respiratory activity. However, it has been shown that mesosomes are not analogous to the mitochondria, since they do not have an outer membrane. Moreover, respiratory enzymes have been shown to occur through the cell membrane.

It is observed that the mesosomal vesicles are either devoid of many of the plasma membrane enzymes and electron transport components (e.g. ATPase, dehydrogenase, cytochromes), or else there are substantially reduced complements of these components. Their function is a matter of conjecture. They have a role in wall synthesis and secretion of extracellular products. It has also been shown that transforming DNA is taken up by all the cells apparently via the mesosomes. The nuclear body gets attached to the mesosome at the time of replication.

Chapter—3 | Classification of Bacteria

Introduction

The classification of bacteria is based on some following specific characteristics: (1) The rigidity of the cell and mode of divisions, (2) Their general shape and type of morphological aggregation, (3) The possession of flagella and their location, (4) Gram staining reaction, (5) The presence or absence of photosynthetic pigments, (6) The presence or absence of sulphur granules, (7) The acid fast nature of the organisms, and (8) The presence of motile spores within the sporangium. These eight characteristics are only the major points in the identification of the bacteria. Physiological activities of the bacteria are also used in the classification. Classical approach, numerical taxonomy chemotaxonomic approach and the most fundamental genetic or molecular approach are several bases for bacterial taxonomy.

Bergey's Manual of Determinative Bacteriology, started in 1923, has been the major treatise on bacterial taxonomy but it could serve only the purpose of identification. It had been considerably changing with respect to composition of taxa higher than genus (families, orders, tribes) in its successive editions. The eighth edition, published in 1974 (Eds. R.E. Buchanan and N.E. Gibbons shows a much modified organisation from the seventh edition. The seventh edition of the manual classifies bacteria in the form of a complete hierarchy. The eighth edition had apodted a more rational approach but the subsequent changes in bacterial classification based on following approaches:

Nature of DNA: It includes the light microscopic observations and staining methods to observe the nature of DNA.

Chemical Composition of the Cell: It includes the biochemical and biophysical techniques to clarify the features of cellular organisation.

Other Cytological Features: The electron microscope has been used to extend the study of comparative cytology to the ultrastructural level of the cell.

Nineteen Sections of Bacterial Group by Bergey's Manual

The eighth edition of the Bergey's Manual had divided the bacteria in 19 sections of which one section is in the division photobacteria while 18 sections are in the second division scotobacteria which contain the true bacteria and mycoplasmas.

Division-I "Photobacteria" includes phototrophic prokaryotes. It has only one section which is further arranged in three classes namely, **Class-I—Bluegreen photobacteria; Class II—Red photobacteria and Class III—Green photobacteria.**

Division-II "Scotobacteria" denotes the prokaryotes indifferent to light. This division is also arranged in three classes namely, **Class-I**—The bacteria having 16 sections, **Class-II—The Rickettsias** having only one section which includes obligate intracellular Scotobacteria in eukaryote cells and **Class III—Mollicutes** also having one section which includes Scotobacteria without cell walls.

The following is a condensed form of the sections with their orders, suborders, family and genus.

Kingdom-Procaryoteae

Division I. Photobacteria (Phototrophic prokaryotes)

Class I	Blue Green Photobacteria	
Class II	Red Photobacteria	} Section I
Class III	Green Photobacteria	

Section I The Phototrophic bacteria—One order with two suborders, three families, 18 Genera and three uncertain affiliation genera.

Division II Scotobacteria (indifferent to light)

Section II The Gliding Bacteria—Two orders with 8 families and 21 genera. There is one genus and two families with 5 genera included under uncertain affiliation.

Section III The sheathed Bacteria—Only 7 genera.

Section IV Budding and/or Appendaged Bacteria—Only 17 genera.

Section V The Spirochetes—One order with one family and 5 genera.

Section VI Spiral and Curved Bacteria—One family with two genera and four unceratain affiliation genera.

Section VII Gram Negative Aerobic Rods and Cocci—Five families with 14 genera and 6 genera of uncertain affiliation

Section VIII Gram Negative Faultatively Anaerobic Rods—Two families with 5 tribes and 17 genera as well as ten genera of uncertain affiliation.

Section IX Gram Negative Anaerobic Bacteria—One family with three genera and 6 genera of uncertain affiliation.

Section X Gram Negative Cocci and Coccobacilli—One family with four genera, two genera of uncertain affiliation.

Section XI Gram Negative Anaerobic Cocci—One family with three genera.

Section XII Gram Negative Chemolithotophic Bacteria—There are three categories of organisms:

(i) Organisms oxidizing ammonia or nitrite. This category has only one family with 7 genera.

(ii) Organisms metabolizing sulphur. This category has only 6 genera.

(iii) Organisms depositing iron or maganese oxides. This category has only one family with four genera.

Section-XIII Methane-Producing Bacteria—One family with three genera.

Section XIV Gram Positive Cocci—Three families with 12 genera.

Section XV Endospore-Forming Rods and Cocci—One family with 5 genera and one genus of uncertain affiliation.

Section XVI Gram Positive Asporogenous Rod-Shaped Bacteria—One family with one genus and three genera of uncertain affiliation

Section XVII Actionomycetes and Related Organisms—One order with 9 families and 33 genera. Four genera of uncertain affiliation. It can be divided into two categories:

(i) Irregular rods—Coryneform group of Bacteria.

(ii) Filamentous Bacteria—Order actinomycetales includes, 8 families.

Section XVIII The Rickettsias—Two orders with four families, three tribes and 18 genera.

Section XIX The Mycoplasmas—One order with two families and two genera and two genera of uncertain affiliation.

An outline of bacterial classification is given below in tabular forms **(Table 1 & 2)** and one another classification is also described after tables.

Table—1: Brief Classification of Bacteria

Kingdom	*Division*	*Class*	*Section*
Procaryotae	1. Photobacteria	1. Blue Green Photobacteria	
	(Phototrophic Prokaryotes)	2. Red Photobacteria	Section-I
		3. Green Photobacteria	
	2. Scotobacteria	1. The Bacteria	Sections-II to XVII
	(Prokaryotes Indifferent to light)	2. The Rickettsias	Section-XVIII
		3. Mollicutes	Section-XIX

Table—2: Detailed Classification of Bacteria

Section	Order	Suborder	Family	Tribe	Genus	Genus (Incertae sedis)
1. The Phototrophic Bacteria	1. Rhodospirillales	1. Rhodospirillineae	1. Rhodospirillaceae	—	1. Rhodospirillum 2. Rhodopseudomonas 3. Rhodomicrobium	1. Chlorochromatium 2. Cylindrogloea 3. Chlorobacterium
			2. Chromatiaceae		1. Chromatium 2. Thiocystis 3. Thiosarcina 4. Thiospirillum 5. Thiocapsa 6. Lamprocystis 7. Thiodictyon 8. Thiopedia 9. Amoebobacter 10. Ectothiorhodospira	
		2. Chlorobilneae	1. Chlorobiaceae	–	1. Chlorobium 2. Prosthecochloris 3. Chloropseudomonas 4. Pelodictyon 5. Clathrochloris	
2. The Gliding Bacteria	1. Myxobacteriales	—	1. Myxococcaceae	—	1. Myxococcus	1. Totothrix
			2. Archangiaceae		1. Archangium	
			3. Cystobacteraceae		1. Cystobacter 2. Melittangium 3. Stigmatella	
			4. Polyangiaceae		1. Polyangium 2. Nannocystis 3. Chondromyces	

(Contd...)

Table—2: (Contd...)

Section	Order	Suborder	Family	Tribe	Genus	Genus (*Incertae sedis*)
	2. Cytophagales	—	1. Cytophagaceae		1. Cytophaga 2. Flexibacter 3. Herpetosiphon 4. Flexithrix 5. Saprospira 6. Sporocytophaga	
			2. Beggiatoaceae		1. Beggiatoa 2. Vitrescilla 3. Thioploca	
			3. Simonsiellaceae		1. Simonsiella 2. Alysiella	
			4. Leucotrichaceae		1. Leucothrix 2. Thiothrix	
			(Familae incertaesedis) 1. Achromatiaceae			1. Achromatium
			2. Pelonemataceae			1. Pelonema 2. Achroonema 3. Peloploca 4. Desmanthos
3. The Sheathed Bacteria	—	—	—	—	1. Sphaerotilus 2. Leptothrix 3. Streptothrix 4. Liesheella 5. Phragmidiothrix 6. Crenothrix 7. Clonothrix	—

Table—2: (Contd...)

Section	*Order*	*Suborder*	*Family*	*Tribe*	*Genus*	*Genus (Incertae sedis)*
4. Budding and /or Appendaged Bacteria	—	—	—	—	1. Hyphomicrobium 2. Hyphomonas 3. Pedomicrobium 4. Caulobacter 5. Asticcacaulis 6. Ancalomicrobium 7. Prosthecomicrobium 8. Thiodendron 9. Pasteuria 10. Blastobacter 11. Seliberia 12. Gallionella 13. Nevskia 14. Planctomyces 15. Metallogenium 16. Caulococcus 17. Kusnezovia	
5. The Spirochetes	1. Spirochaetales	—	—	—	1. Spirochaeta 2. Cristispira 3. Treponema 4. Borrelia 5. Leptospira	
6. Spiral and Curved Bacteria	—	—	1. Spirillaceae	—	1. Spirillum 2. Campylobacter	1. Bdellovibrio 2. Microcyslus 3. Pelosigma 4. Brachyarcus

(Contd...)

Table—2: (Contd...)

Section	Order	Suborder	Family	Tribe	Genus	Genus (Incertae sedis)
7. Gram Negative Aerobic Rods and Cocci	—	—	1. Pseudomonadaceae	—	1. Pseudomonas 2. Xanthomonas 3. Zoogloea 4. Gluconobacter	1. Alcaligenes 2. Acetobacter 3. Brucella 4. Bordetella 5. Francisella 6. Thermus
			2. Azotobacteraceae	—	1. Azotobacter 2. Azomonas 3. Beijerinckia 4. Derxia	
			3. Rhizobiaceae	—	1. Rhizobium 2. Agrobacterium	
			4. Methylomonadaceae		1. Methylomonas 2. Methylococcus	
			5. Halobacteriaceae	—	1. Helobacterium 2. Halococcus	
8. Gram Negative Facultatively Anaerobic Rods	—	—	1. Enterobacteriaceae	1. Escher-ichiae	1. Escherichia 2. Edwardsiella 3. Citrobacter 4. Salmonella 5. Shigella	1. Zymomonas 2. Chromobacterium 3. Flavobacterium 4. Haemophilus 5. Pasteurella 6. Actinobacillus 7. Cardiobacterium 8. Neguchia
				2. Klebsi-ellae	1. Klebsiella 2. Enterobacter 3. Hafnia 4. Serratia	

Table 2:—(Contd...)

Section	*Order*	*Suborder*	*Family*	*Tribe*	*Genus*	*Genus (Incertae sedis)*
				3. Proteae	1. Proteus	
				4. Yersini-eae		1. Yersinia
				5. Erwin-ieae	1. Erwinia	
			2. Vibrionaceae		1. Vibrio 2. Aeromonas 3. Plesiomonas 4. Photobacterium 5. Lucibacterium	
9. Gram Negative Anaerobic Bacteria	—	—	1. Bacteriodaceae	—	1. Bacteroides 2. Fusobacterium 3. Leptotrichia	1. Desulfovibrio 2. Butyrivibrio 3. Succinivibrio 4. Succinimonas 5. Lachmospira 6. Selenomonas
10. Gram Negative Cocci and Cocco-bacilli	—	—	1. Neisseriaceae	—	1. Neisseria 2. Branhamella 3. Moraxella 4. Acinetobacter	1. Paracoccus 2. Lampropedia
11. Gram Negative Anaerobic cocci	—	—	1. Veillonellaceae	—	1. Veillonella 2. Acidamino coccus 3. Megasphaera	—
12. Gram Negative Chemolithotophic Bacteria	—	—	1. Nitrobacteraceae (Ammmonia or nitrite oxidzing organisms)		1. Nitrobacter 2. Nitrospina 3. Nitrococcus 4. Nitrosomonas 5. Nitrosospira 6. Nitrosococcus .7. Nitrosolobus	

(Contd...)

Table—2: (Contd...)

Section	Order	Suborder	Family	Tribe	Genus	Genus (Incertae sedis)
			(Sulphur metabolizing organisms)		1. Thiobacillus 2. Sulfolobus 3. Thiobacterium 4. Macromonas 5. Thiovulum 6. Thiospira	
			2. Siderocapsaceae (Iron or manganese oxides depositing organisms)		1. Siderocapsa 2. Naumanniella 3. Ochrobium 4. Siderococcus	
13. Methane Producing Bacteria			1. Methanobacteriaceae		1. Methenobacterium 2. Methenosarcina 3. Methanococcus	
14. Gram Positive Cocci			1. Micrococcaceae		1. Micrococcus 2. Staphylococcus 3. Planococcus	
			2. Streptococcaceae		1. Streptococcus 2. Leuconostoc 3. Pediococcus 4. Aerococcus 5. Gemella	
			3. Peptococcaceae		1. Peptococcus 2. Peptostreptococcus 3. Ruminococcus 4. Sarcina	
15. Endospore Forming Rods and Cocci			1. Bacilaceae		1. Bacillus 2. Sporolactobacillus 3. Clostridium 4. Desulfotomaculum 5. Sporosarcina	1. Oscillospira

Table—2: (Contd...)

Section	Order	Suborder	Family	Tribe	Genus	Genus (Incertae sedis)
16. Gram Positive, Asporogenous Rod Shaped Bacteria	—	—	1. Lactobacillaceae	—	1. Lactobacillus	1. Listeria 2. Erysipelothrix 3. Caryophanon
17. Actinomycetes and Related organisms	(Coryneform group of Bacteria	—	—	—	1. Corynebacterium 2. Arthrobacter	1. Brevibacterium 2. Microbacterium 3. Cellulomonas 4. Kurthia
			1 Propionibacteriaceae		1. Propionibacterium 2. Eubacterium	
	1. Actinomycetales	—	1. Actinomycetaceae	—	1. Actinomyces 2. Arachnia 3. Bifidobacterium 4. Bacterionema 5. Rothia	
			2. Mycobacteriaceae	—	1. Mycobacterium	
			3. Frankiaceae	—	1. Frankia	
			4. Actionoplanaceae	—	1. Actinoplanes	

(Contd...)

Table—2: (Contd...)

Section	*Order*	*Suborder*	*Family*	*Tribe*	*Genus*	*Genus (Incertae sedis)*
					2. Spirillospora	
					3. Streptosporangium	
					4. Amorphosporangium	
					5. Ampullariella	
					6. Pilimelia	
					7. Planomonospora	
					8. Planobispora	
					9. Dactylosporangium	
					10. Kitasatoa	
			5. Dermatophilaceae	—	1. Dermatophilus	
					2. Geodermatophilus	
			6. Nocardiaceae	—	1. Nocardia	
					2. Pseudonocardia	
			7. Streptomy cetaceae	—	1. Streptomyces	
					2. Streptoverticillum	
					3. Sporichthya	
					4. Microellobosporia	
			8. Micromono sporaceae	—	1. Micromonospora	
					2. Thermoactinomyces	
					3. Actinobifida	
					4. Thermomonospora	
					5. Microbispora	
					6. Micropolyspora	

Table—2: (Contd...)

Section	Order	Suborder	Family	Tribe	Genus	Genus (Incertae sedis)
18. The Rickettsias	1. Rickettsiales		1. Rickettsaceae	1. Rickett-sieae	1. Rickettsia 2. Rochalimaea 3. Coxiella	
				2. Ehrli-chieae	1. Ehrlichia 2. Cowdria 3. Neorickettsia	
				3. Wolba-chieae	1. Wolbachia 2. Symbiotes 3. Blattabacterium 4. Rickettsiella	
			2. Bartonellaceae	—	1. Bartonella 2. Grahamella	
			3. Anaplasmataceae		1. Anaplasma 2. Paranplasma 3. Aegyptianella 4. Haemobartonella 5. Eperythrozoon	
	2. Chlamydiales	—	1. Chlamydiaceae	—	1. Chlamydia	
19. The Mycoplasmas	1. Mycoplasmatales	—	1. Mycoplasmataceae 2. Acholeplasmataceae		1. Mycoplasma 1. Acholeplasma	1. Thermoplasma 2. Spiroplasma

Groups of Bacteria

Phylum 1 Proteobacteria

I. Purple Phototrophic Bacteria

(i) **Gram-negative** rods (1.5-4 × 4-40 μm), spirals, ovoid, bean-shaped; some rods motile via polar flagella

(ii) **Strictly Anaerobic Anoxygenic Phototrophs** that use the **Calvin Cycle** for CO_2 fixation; have bacteriochlorophylls and use reduced molecules such as hydrogen sulfide, sulfur, thiosulfate or hydrogen as electron source for generation of NADH and NADPH:

(a) **Purple Sulfur Bacteria:** Chromatium, Halorhodospira, Thiocapsa, Thiococcus, Thiopedia,Thiospirillum are photolithoautotrophs and often form sulfur granules inside their cells

(b) **Purple Nonsulfur Bacteria:** Rhodobacter, Rhodospirilum,Rhodocyclus, Rhodobacter Rhodopseudomonas, Rhodophila are photo-organotrophs

(iii) **Found in Anaerobic, Sulfide-rich Zones** of lakes and lake muds

II. Nitrifying Bacteria—Nitrobacter, Nitrosomonas

(i) **Gram-negative** rods (0.8-1 × 1-2 μm), coccoid (0.2-1 μm), spiral (0.3-0.4 μm in diameter), lobular (1-1.5 μm); may have extensive membrane complexes in cytoplasm

(ii) **Aerobic Lithotrophs** which use carbon dioxide or carbonate as carbon source (via Calvin cycle) and derive energy by oxidizing inorganic compounds containing reduced nitrogen (NH4+, nitrite-)

(iii) **Found in Soil, Sewage Treatment Systems, Freshwater, and Marine Habitats**

III. Sulfur and **Iron-Oxidizing Bacteria**—Thiobacillus, Beggiatoa, Thioploca, Thiothrix

(i) **Gram-negative** rods (0.8-1 × 1-2 μm), coccoid (0.2-1 μm), spiral (0.3-0.4 μm in diameter), lobular (1-1.5 μm); may have extensive membrane complexes in cytoplasm; or filamentous with gliding motility (Beggiatoa)

(ii) **Aerobic Lithotrophs** use carbon dioxide or carbonate as carbon source (via Calvin cycle) and derive energy by oxidizing inorganic compounds containing reduced sulfur (Thiobacillus, Thiomicrospora) and/or reduced sulfur and iron (Thiobacillus ferroxidans)

(iii) **Found in Soil, Sewage Treatment Systems, Freshwater, and Marine Habitats, Especially Sulfur Springs**

IV. Hydrogen-oxidizing Bacteria—Alcaligenes

(i) **Gram-negative** rods (0.8-1 × 1-2 μm), coccoid (0.2-1 μm), spiral (0.3-0.4 μm in diameter), lobular (1-1.5 μm); may have extensive membrane complexes in cytoplasm

(ii) **Aerobic Lithotrophs** which use carbon dioxide or carbonate as carbon source (via Calvin cycle) and derive energy by oxidizing inorganic compounds containing reduced hydrogen (generally hydrogen)

(iii) **Found in Soil, Sewage Treatment Systems, Freshwater, and Marine Habitats**

V. Methanotrophs and Methyltrophs—Methlyosinus, Methylcoccus

(i) **Gram-negative** motile rods (Methylmonas), nonmotile cocci (Methylcoccus) or motile vibrios (Methylosinus); possess sterols

(ii) **Aerobic Organotrophs** generate energy by oxidizing methane (or methyl groups)

(iii) **Widespread** in nature in soil and water

VI. Pseudomonas and the Pseudomonads—Pseudomonas, Agrobacterium, Rhizobium, Zymomonas

(i) **Gram-negative** straight or slightly curved rods (0.5 × 1.5 to 1.4 × 6 μm); motile (polar flagella)

(ii) **Aerobic and Facultative Aerobic Organotrophs**; respiratory (non-fermentative) metabolism that utilizes the Entner-Doudoroff pathway for carbohydrate oxidation (Zymomonas uses the Entner-Doudoroff pathway for oxidation of glucose to ethanol); versatile use of carbon and energy sources, including use of nitrate as final electron acceptor via **anaerobic respiration**; Rhizobia are **aerobic** organisms which live inside other organisms

(iii) **Free-living to Parasitic in Animals** (Pseudomonas) **and plants** (Pseudomonas, Xanthomonas); Rhizobium is a plant symbiote; Zymomonas is **found in soil and water;** occasionally present in pyogenic infections of humans and animals

VII Acetic Acid Bacteria (Acetobacteraceae)—Acetobacter, Gluconobacter

(i) **Gram-negative** straight or curved rods and cocci (0.3 × 2 to 0.9 × 20 μm); motile - polar (Gluconobacter) or peritrichous (Acetobacter) flagella

(ii) **Aerobic** respiratory **organotrophs**; derive energy by oxidizing ethanol to acetic acid

(iii) **Free-living**, important in the food industry as vinegar producers

VIII. Free-Living Aerobic Nitrogen-Fixing Bacteria—**Azotobacter,** Azomonas

(i) **Gram-negative**, large, rod to pear-shaped (2-4 μm in diameter)

(ii) **Aerobic or Microaerophilic Organotrophic** nitrogen-fixers; use organic and amino acids as carbon and energy source, not carbohydrate;

Azotobacter has fastest rate of oxygen-uptake of any known organism

(iii) **Free-living Terrestrial Nitrogen-fixers**, including Azospirillum, Azotobacter, Azomonas

IX. Neisseria, Chromobacterium, and Relatives—Neisseria, Chromobacterium

(i) **Gram-negative** cocci (Neisseria) or rod-coccoid; generally nonmotile (Moraxella and Acinetobacter possess **twitching motility**); Chromobacterium violaceum has bright purple pigment

(ii) **Aerobic Respiratory** (non-fermentative) **Organotrophs**; oxidase positive; versatile carbon and energy sources; Chromobacterium ferments a variety of carbohydrates and produces a violet-coloured pigment (violacein) when grown in the presence of tryptophan

(iii) **Isolated from Animals** (Neisseria are generally parasitic, some Branhamella are symbiotic with humans) **or from soil and water** (Acinetobacter); Chromobacterium and Acinetobacter are occasionally present in pyogenic infections of humans and animals

X. Enteric Bacteria—Escherichia, Salmonella, Proteus, Enterobacter

(i) **Gram-negative,** straight rods (0.3-1 × 1-6 μm); nonmotile to motile (peritrichous flagella)

(ii) **Facultative Aerobes;** oxidase negative; diverse (including fermentative) **organotrophic** catabolism

(iii) **Normal Flora and/or Parasitic** on mammals, birds and plants

XI. Vibrio and Photobacterium—Vibrio, Photobacterium

(i) **Gram-negative** curved rods (0.3-1 × 1-6 μm); nonmotile to motile (peritrichous or polar flagella)

(ii) **Facultative Aerobes**: oxidase positive; diverse (fermentative) organotrophic catabolism

(iii) **Typically Aquatic** (freshwater or marine) **free-living;** Vibrio cholerae and Vibrio parahemolyticus are parasitic on animals, especially humans; Photobacterium is bioluminescent

XII. Rickettsias—Coxiella, Rickettsia, Rochalimaea

(i) **Gram-negative** rod, coccoid, pleomorphic (0.3-0.7 × 1-2 μm); nonmotile

(ii) **Aerobic Organotrophs:** Coxiella oxidize glucose; Rickettsia do not oxidize glucose or organic acids, but obtain ATP from oxidation of glutamate or glutamine, which they obtain from the host cell (there is some suggestion that they also obtain NAD+ and coenzyme A from the host cell), and they replicate in the cytoplasm of the host cell until it is loaded with parasites

(iii) **Pathogens—transmitted to humans by** arthropods; Rickettsia are obligate intracellular parasites; Rochalimaea grow attached to outside of host cell cytoplasmic membrane

XIII. Spirilla—Spirilum, Bdellovibrio, Campylobacter

(i) **Gram-negative**; helical to vibrioid (curved = helical but not long enough to encompass one full helical turn; 0.2 × 1.4 to 1.7 × 60 μm); motile via flagella (Spirillum, Aquaspirillum) or nonmotile (Spirosomaceae, Microcyclus)

(ii) **Aerobic, Microaerophilic or Facultative Aerobic Organotrophs**; use organic and amino acids as carbon/energy source, not carbohydrate

(iii) **Free-living** (freshwater - Spirillum, Aquaspirillum; marine - Oceanospirillum), **soil** (some nitrogen-fixers - Azospirillum), parasitic (animals—Campylobacter; bacteria - Bdellovibrio)

XIV. Sheathed Proteobacteria—Sphaerotilus, Leptothrix

(i) **Gram-negative**; filamentous—chains of cells surrounded by hollow tubelike structure (sheath) used for surface attachment (frequently have

holdfasts), nutrient adsorption and protection from predators; unicellular elements (swarmers) generally motile by subpolar flagella, some nonmotile (Crenothrix)

(ii) **Aerobic Metal Oxidizers:** precipitate iron or manganese on sheath; Lepthothrix oxidizes manganese, Sphaerotilus doesn't

(iii) **Freshwater (flowing) Habitats; Polluted Streams and Activated Sludge**

XV. Budding and Prosthecate/Stalked Bacteria—Hyphomicrobium, Caulobacter

(i) **Gram-negative** rods (0.5-1 × 1-3 μm); mobile by flagella; heterogeneous group, but reproduce by budding and/or possess a prostheca (cytoplasmic extrusions such as stalks, hyphae, or other appendages); motile by flagella; Hyphomicrobium - stalks and budding; Caulobacter-stalk with holdfast; reproduces by budding

(ii) **Generally Aerobic Organotrophs** (Rhodomicrobium is an exception ... it is phototrophic)

(iii) frequently found in **nutrient-poor freshwater habitats**

XVI. Gliding Myxoacteria—Myxococcus, Stigmatella

(i) **Gram-negative** rods (0.6-0.9 × 3-8 μm) with gliding motility and production of fruiting bodies (50-500 μm tall) which contain myxospores (Myxococcus, Chondromyces)

(ii) **Aerobic** respiratory **Organotrophs:** micropredators which kill other bacteria with antibiotics, then secrete enzymes to dissolve them

(iii) **Found Worldwide in Soil, Decaying Plant Material and Animal Dung**; most abundant in warm areas; also found in arctic tundra

XVII. Sulfate and Sulfur-Reducing Bacteria—Desulfovibrio, Desulfobacter, Desulfuromonas *(i)* **Gram-negative** (although some are Gram-positive endospore formers); straight, curved, or helical rods (0.5-1.5 × 3-10 μm)

(ii) **Anaerobic Dissimilatory Organotrophic** bacteria which use sulfate or sulfur as electron acceptors in anaerobic respiration of organic compounds such as acetate, which they oxidize via an acetyl-CoA pathway; when they have a source of hydrogen and carbon dioxide, but no organic compounds, they can use hydrogenase to generate energy during lithotrophic growth

(iii) thrive in **fresh water sediments and sewage digestors**; important in sulfur cycle; combat acid rain

Phylum 2 Gram-Positive Bacteria

I. Nonsporulating, Low GC, Gram-Positive Bacteria (plus Lactic Acid Bacteria)—Staphylococcus, Micrococcus, Streptococcus, Lactobacillus

(i) **Gram-positive**, nonmotile **cocci** (0.2-2.5 μm), single or in variable-sized clusters (Staphylococcus, Micrococcus) or in chains (Streptococcus) **or Gram-positive**, usually nonmotile **bacilli** (0.8 × 2 μm - Lactobacillus) single or in variable-sized clusters; nonsporulating

(ii) **Aerobic Organotrophs** (Micrococcus); **Facultatively Aerobic or Microaerophilic Aerobic Organotrophs** (Staphylococcus, Streptococcus, Leuconostoc); anaerobic organotrophs (Lactobacillus) that lack porphyrins (no cytochromes) so must obtain energy exclusively by substrate-level phosphorylation; **lactic acid bacteria are:**

(a) **Homofermentative**—produce lactic acid as sole product of fermentative metabolism (Streptococcus, Pediococcus, Enterococcus, Lactococcus, some Lactobacillus species)

(b) **Heterofermentative**—produce other products (ethanol, etc.) in addition to lactic acid (Leuconostoc, some Lactobacillus species)

(ii) **Free in Nature, Normal Flora, Parasites** (Staphylococcus, Streptococcus); many lactic acid bacteria (Lactobacillus, Streptococcus, Leuconostoc) are useful in food production because of the lactic acid they generate

II. Endospore-Forming Gram-Positive Rods and Cocci (plus Homoacetogenic Bacteria and Heliobacteria)—Bacillus, Clostridium, Sporosarcina, Heliobacterium

(i) **Gram-positive** endosporeforming bacilli (0.3-2.2 × 1.2-6 μm); usually motile by peritrichous flagella (Bacillus, Clostridium) or Gram-positive gliding rod and motile spirilla (Heliobacterium—two species produce endospores) or actively motile rods (Heliobacillus) or cocci in tetrads or packets of 8 or more cells (Sporocarcina)

(ii) **Physiology is Highly Variable**

(a) **Facultatively Aerobic** (Bacillus)

(b) **Microaerophilic** (Sporolactobacillus); **anaerobic** (Clostridium, Desulfotomaculum, Sporosarcina - lack catalase)

(iii) **Obligately Anaerobic Organotrophs** produce acetate as the sole product of sugar fermentation (to produce 2 acetates plus $2CO_2$ and H_2) coupled with reduction of carbon dioxide (via the acetyl-CoA pathway) to form a third acetate thus the name homoacetogenic (the sole or only product of sugar catabolism is acetate); may also grow lithotrophically using hydrogen as an energy source (same acetyl-CoA pathway as above) (Acetobacterium, Clostridium, Desulfotomaculum)

(iv) **Obligately Anaerobic Phototrophs** (have bacteriochlorophyll g), but can utilize acetate,

pyruvate, lactate or butyrate as carbon sources for **organotrophic** growth (Heliobacterium)

(v) **Ubiquitous in Nature** (soil); Heliobacterium is found in tropical soils, especially rice paddies; Bacillus and Clostridium are important in food spoilage, infectious disease

III. Cell Wall-Less, Low GC, Gram-Positive Bacteria: The Mycoplasmas—Mycoplasma, Spiroplasma

(i) **Gram-positive** pleomorphic (0.1-0.25 × 3-150 μm); lack cell wall, possess sterols; most are non-motile, some have **gliding motility**

(ii) **Facultatively Aerobic or Obligately Anaerobic Organotrophs**; Mycoplasmataceae require steroids, Acholeplasmataceae do not

(iii) **Ubiquitous Parasites** (may have arisen from Clostridium or Gram-positive Archaea)

V. High GC, Gram-Positive Bacteria: Coryneform and Propionic Acid Bacteria—Corynebacterium, Arthrobacter, Propionibacterium

(i) **Gram-positive**; nonmotile straight or slightly curved **bacilli**, sometimes with swellings (club-shaped Coryneforms); or **Gram-positive** pleomorphic, nonsporulating **bacilli** (0.8 × 2 μm Propionibacterium)

(ii) **Aerobic Organotrophs** (Coryneforms); **Anaerobic, Organotrophic** versatile fermenters (Propionibacterium)

(iii) **Animal or Plant Pathogens or Normal Microbiota** (Corynebacterium) or **soil saprophytes** which mineralize soil contaminants (Arthrobacter); Propionibacterium grows **free in nature, normal flora** and is useful in food production (swiss cheese)

V. High GC, Gram-Positive Bacteria: Mycobacterium

(i) **Gram-positive** (weakly), slightly curved or straight rods (0.2-0.6 × 1-10 μm); cell walls have

long-chain waxes containing mycolic acids (acid-fast)

(ii) **Aerobic Organotrophs**; growth slow or very slow

(iii) important as **pathogens**

VI. Filamentous, High GC, Gram-Positive Bacteria: The Actinomycetes—Streptomyces, Actinomyces

(i) **Nocardioforms**

(a) **Gram-variable:** Nocardia develop a highly branched substrate mycelium (hyphae 0.5-1.2 µm in diameter) that breaks up into rods and cocci, form aerial hyphae, and produce conidia; others may (Nocardioides) or may not (Rhodococcus) form aerial hyphae or produce conidia

(b) **Mesophilic, Aerobic Organotrophs:** Nocardia degrades hydrocarbons

(c) **World-wide in Soil and Water:** mostly free-living; Nocardia is pathogenic

(ii) **Actinomycetes with Multilocular Sporangia**

(a) **Gram-positive;** 0.5-2 µm diameter hyphae; clusters of coccoid spores, some motile via tufts of flagella (Dermatophilus), others non-motile (Frankia)

(b) **Facultatively Anaerobic or Microaerophilic Organotrophs** (non-fermentative)

(c) Dermatophilus is **parasitic,** Causes streptothrichosis; Frankia is symbiotic with roots of non-leguminous plants (Alder trees) and fixes nitrogen

(iii) **Actinoplanetes**

(a) **Gram-positive,** extensive branched, non-fragmenting substrate mycelium with 0.2-2.6 µm diameter hyphae; highly coloured; motile spores formed in sporangia (3-20 ×

6-30 μm, spherical-cylindrical, very irregular) on sporangiophores

(b) **Mesophilic Aerobic Organotrophic** decomposers

(c) **Found in Soil** (forest litter to beach sand), freshwater and marine habitats; important in mineralization of organics (Pilimelia) and production of gentamicin (Micromonospora)

(iv) **Streptomyces and Related Genera**

(a) **Gram-positive; a**erial hyphae (0.5-2 μm diameter) which form chains of motile conidio-spores; mycelium does not undergo fragmentation; variety of colours

(b) **Aerobic** mesophiles with flexible organotrophic metabolism; degrade chitin, cellulose, pectin, keratin, etc.

(c) **Abundant Soil Microbes** (up to 20% of total) which are responsible for the odor of soil (geosmin) and are important in mineralization; Streptomyces also produce antibiotics (antifungals—amphotericin B and nystatin; ribosome active antibacterials—streptomycin, neomycin, erythromycin, tetracycline, chloramphenicol); some cause diseases such as actinomycetoma

(v) **Maduromycetes**

(a) **Gram-positive;** aerial mycelia with non-fragmenting mycelium which forms pairs or short chains of arthrospores; some form sporangia

(b) **Aerobic Organotrophs:** most mesophilic, some thermophilic

(c) **Parasitic** — Actinomadura causes actinomycetomas in humans

(vi) **Thermonospora and Related Genera**

(a) **Gram-positive;** variable morphology; single heat-sensitive spores on aerial mycelium and/or substrate mycelium

(b) **Aerobic** thermophilic (40-48°C) organotrophs

(c) **High Temperature Habitats** (compost piles, haystacks, etc.)

(vii) **Thermoactinomycetes**

(a) Gram-positive; hyphae (0.4-0.8 μm diameter) in substrate mycelium; single endospores (heat resistant) on both aerial and substrate mycelia

(b) aerobic thermophilic (45-60°C) organotrophs

(c) Thermoactinomyces vulgaris is one of the causative agents of farmer's lung disease (allergic disease of respiratory system)

(viii) **Other Genera**

(a) **"Catch-all" Category** for Actinomycete genera whose characteristics are different from those found above

(b) **Current Members** include: Glycomyces, Kibdelosporangium, Kitasatosporia, Saccharothrix, Pasteuria

Phylum 3 Cyanobacteria, Prochlorophytes, and Chloroplasts

I. Cyanobacteria—Synechococcus, Oscillatoria, Nostoc

(i) **Gram-negative** 1-10 μm diameter, unicellular (Anacystis, Synechococcus, Pleurocapsa) or filamentous (Anabena, Nostoc, Oscillatoria); gliding motility; some have gas vesicles for vertical movement in water

(ii) **Aerobic Oxygenic Photolithotrophs** with chlorophyll a plus phycobiliproteins which use water as electron source for generation of NADH and NADPH and generate oxygen; heterocystous cyanobacteria (Anabena, Nostoc) have heterocysts which facilitate nitrogen-fixation

(iii) **Responsible for Most of the Oxygen in the Atmosphere and for Some of the Nitrogen-Fixation**; very good at symbiotic relationships

(e.g., the "algae" in most lichens are Cyanobacteria); ecology, etc.

II. Prochlorophytes and Chloroplasts—Prochloron, Prochlorothrix

(i) **Gram-negative**; 8-10 µm diameter, spherical (Prochloron) with extensive thylakoid membrane system (because it closely resembles them, and is frequently endosymbiotic, some think Prochloron is the precursor of chloroplasts); filamentous (Prochlorothrix) with meager thylakoid membrane development

(ii) **Aerobic Oxygenic Photolithotrophs** with chlorophyll a or b (but lack phycobiliproteins); picoplankton is ~1 µm in diameter and has only chlorophyll b)

(iii) **Endosymbionts of Marine Invertebrates**

Phylum 4 Chlamydias—Chlamydia

(i) **Gram-negative** coccoid (0.2-1.5 µm); non-motile; no peptidoglycan

(ii) **Aerobic Organotrophs:** most limited catabolic and biosynthetic capabilities of any known organism; no energy-generating system—obtain ATP and metabolic intermediates from host cell (energy parasites); replicate via binary fission of reticulate bodies, which then differentiate into small, dense cell forms called elementary bodies, which are specialized for transmission to a new host

(iii) **Pathogens**—obligate intracellular parasites; life cycle involves elementary and reticulate bodies)

Phylum 5 Planctomyces/Pirella—Planctomyces, Pirella

(i) **Gram-negative** ovoid bacteria with a slender stalk, pili and a flagellum; cell wall contains protein instead of peptidoglycan; reproduces by

budding and expresses a life cycle; cells include membrane-bound structures, only some of which contain the genome (Gemmata obscuriglobus)

(ii) **Facultatively Aerobic Organotrophs that Ferment or Respire Sugars**

(iii) **Aquatic Habitats** freshwater or marine

Phylum 6 Verrucomicrobia—Verrucomicrobium, Prosthecobacter

(i) **Gram-negative** with peptidoglycan with several prosthecae per cell (verucco = "warty")

(ii) **Aerobic** to **Facultatively Aerobic Ferment Sugars**

(iii) Found in **freshwater/marine waters and forest/agricultural soils**

Phylum 7 Bacteroides/Flavobacteria—Bacteroides, Flavobacterium

(i) **Gram-negative** straight, curved or helical rods (0.5 × 0.5-10 μm); nonmotile or motile via flagella (Bacteroides possess sphingolipids) or yellow pigmented rods (Flavobacterium) that move by **gliding motility** utilizing movement of proteins on cell surface

(ii) **Anaerobic Organotrophs:** Bacteroides ferment many sugars to acetate and succinate, whereas Flavobacteria ferment glucose (exclusively)

(iii) **Commensals** in the oral cavity and/or intestinal tract of humans and other animals, rumen of cattle, etc. (Bacteroides); freshwater and marine aquatic environments, and found in foods and food-processing plants (Flavobacterium)

Phylum 8 Cytophaga—Cytophaga, Sporocytophaga, Flexibacter

(i) **Cytophaga, Sporocytophaga**

(i) **Gram-negative** long, slender rods (often with pointed ends) that move by gliding motility

(ii) **Obligately Aerobic Organotrophs** that **Digest Cellulose** with cellulases that remain bound to the cell envelope (not secreted)

(iii) found on surfaces of the cellulose-containing material (or fish gills)

II. Flexibacter

(i) **Gram-negative** filaments that move by gliding motility; many possess carotenoids

(ii) **Obligately Aerobic Organotrophs** that **do not** digest cellulose require complex media

(iii) Common **soil and freshwater saprophytes**

Phylum 9 Green Sulfur Bacteria—Chlorobium, Prosthechloris, Chlorochromatium

(i) **Gram-negative** nonmotile rods, cocci or spirals (0.5-1.1 μm wide or diameter) often form sulfur granules outside their cells

(ii) **Obligately Anaerobic Photolithotrophs** which have bacteriochlorophylls in chlorosomes and use reduced molecules such as hydrogen sulfide, sulfur, thiosulfate or hydrogen as electron source for generation of NADH and NADPH

(iii) **Found in Anaerobic, Sulfide-rich Zones** of lakes and lake muds

Phylum 10 Spirochetes—Spirochaeta, Treponema, Cristispira, Leptospira, Borrelia

(i) **Gram-negative**; helical, long and slender (0.1-5 x 0.75-250 μm); motile via axial filaments

(ii) **Aerobic, Microaerophilic, Facultatively Aerobic, or Anaerobic Organotrophs**; catabolize carbohydrates, amino acids or lipids

(iii) **Free-living** (Spirochaeta), symbiotic (Cristispira), or parasitic (Treponema, Borrelia, Leptospira)

Phylum 11 Deinococci—Dienococcus, Thermus

(i) **Gram-Positive** but Deinococcus has an outer membrane and is not phylogenetically related to other Gram-positive bacteria

(ii) **Aerobic Organotroph** that is highly resistant to radiation damage; the best-known representative of the thermophilic genus Thermus is Thermus aquaticus, the bacterium from which Taq polymerase is derived

(iii) **Free in Nature**, especially near radioactive materials (Dienococcus) or in hot springs (Thermus)

Phylum 12 Green Nonsulfur Bacteria—Chloroflexus, Thermomicrobium

(i) **Gram-negative** motile (gliding) filamentous bacteria

(ii) **Thermophilic Obligately Anaerobic Phototrophs** (Chloroflexus) which have bacteriochlorophylls in chlorosomes and use reduced molecules such as hydrogen sulfide, sulfur, thiosulfate or hydrogen as electron source for generation of NADH and NADPH, but can also grow lithotrophically or organotrophically under appropriate conditions; or thermophilic aerobic organotrophs (Thermomicrobium) with no glycerol and neither ester- nor ether-linkages in their membrane lipids

(ii) **Found in Anaerobic, Sulfide-rich Zones** of lakes and lake muds

Phylum 13 Thermotoga and Thermodesulfobacterium

I. Thermotoga

(i) **Gram-negative** rod with a sheath-like envelope (hence the term "toga" in the name)

(ii) **Thermophilic Anaerobic Fermenters (organotrophic)**

(iii) Grows around benthic **hydrothermal vents** and continental **hot springs**

II. Thermodesulfobacterium

(i) **Gram-negative** bacillus with **ether-linked** (non-phytanyl) **lipids** (very Archaea-like!)

(ii) **Strictly Anaerobic Thermophilic Sulfate-reducing Organotroph** with an optimum growth temperature of 70°C (most thermophilic of all known sulfate-reducing bacteria)

(iii) Grows in high-temperature areas near **hot springs**, etc.

Phylum 14 Aquifex, Thermocrinus and Relatives

(i) **Gram-negative** bacillus

(ii) **Hyperthermophilic** (optimum temperature for growth is 85°C, but can grow up to 95°C most thermophilic of all known bacteria) **aerobic** (one of the few aerobic hyperthermophiles known) **lithotroph** (reverse TCA cycle) that oxidizes hydrogen, sulfur or thiosulfate and uses oxygen (microaerophilic growth) or nitrate (anaerobic growth) as terminal electron acceptors

(iii) **Submarine volcanic hot spring bacterium**; most ancient branch of the bacterial phylogenetic tree

Phylum 15 Nitrospira Group

(i) Nitrospira—Gram-negative, aerobic oxidizer (NO_2^- NO_3^-)

(ii) Leptospirillum—Gram-negative, aerobic oxidizer (Fe^{2+} Fe^{3+})

(iii) Thermodesulfovibrio - Gram-negative thermophilic anaerobic sulfate reducer

Phylum 16 Deferribacter group—Gram-negative anaerobes

(i) Defferibacter, Geovibrio - anaerobic respirers that reduce Fe^{3+} and or Mn^{2+}

(ii) Flexistipes - fermenter

Bacterial Nomenclature

In spite of the existence of strict rules for bacterial nomenclature the establishment of new species had been on "new host-new species" concept basis and this had resulted in multiplicity of names given to bacteria that were basically the same. Sometimes, there is found a confusion that some bacterial species capable of producing different symptoms on different hosts or variants of the same bacterium producing different symptoms on the same hosts. The International Committee on Systematic Bacteriology (ICSB) took certain steps for solving this confusing problem. The International Code on Bacterial Nomenclature was revised in 1976. This code provided that on January 1, 1980, all names (class, order, tribe, family, genera and species) published prior to this date and included in the "Approved Lists of Bacterial Names" (Skerrman, et al., 1980) shall be treated as though they had been validity published for the first time on that date. Those names considered valid prior to January 1, 1980 but not induded in the "Approved lists" will have no further standing in bacterial nomenclature and such names can be used for other taxa proposed in future.

Chapter—4 Special DNA Characteristics of Bacteria

Introduction

Under the laboratory microscope the hereditary material of bacteria appear as small and round or oval bodies, devoid of any membrane. Under the electron microscope, loosely packed 20Å fibres of double-stranded DNA are visible in these apparently amorphous structures. The bacterial chromosomal DNA is located in a region of the cell known as the nucleoid. Bacteria, being prokaryotes, do not have a true, membrane bound nucleus. Bacteria carry a single chromosome that is circular in structure. Additional genetic information may be carried on plasmids. These are circles of DNA that lie within the bacterial cytoplasm and replicate independently of the chromosome. Plasmids carry genes that are typically not essential for survival, but that can confer selective advantages in special circumstances. Not all bacterial cells carry plasmids, but some can carry several plasmids in a single cell. R-factors are plasmids that carry genes that confer antibiotic resistance on the cell. Toxins are sometimes coded for by plasmid genes. Sexuality in bacteria is controlled by cytoplasmic factor which is known as sex factor or F-factor. Female or F^- cells lack the sexfactor. Male or F^+ cells contain the sex factor and transfer of genetic material takes place from F^+ To F^- cells. For genetic transfer it is necessary that the sex factor is integrated into the bacterial chromosome. Such F^+ cells are known as HFr because they can transfer the genetic material with high frequency. Thus, depending on whether the F-factor is free or integrated into the chromosome, bacterial males can be F^+ or Hfr. In the integrated state the F-factor or episome multiplies or replicates along with the bacterial chromosome. In the cytoplasmic state it has the capacity to multiply and replicate independently of the bacterial chromosome. Episomes are infectious too, i.e. they can come out of F^+ cells and can enter other cell. Episomes have been shown to be nothing but a circular piece of DNA. Integration of circular episome into the circular bacterial chromosome occurs as a result of recombination.

Lysogenes are bacteria that have been stably infected with a bacteriophage and that carry the virus as a ' prophage'. The bacteriophage DNA is integrated into the genome of the bacterium. Under special conditions, lysogens can burst to release new bacteriophage particles.

Hereditary material of bacteria is the most vital part of the cell and contains the information for various characters of the organisms. Nucleiod, Plasmids and F-factor etc. are some special DNA features that give the speciality to the bacterial hereditary material. The details of these specific DNA parts are given below:

Nucleoid: As previously celared that the genetical material of bacteria is a skein of circular DNA localised as the nucleoid. The nucleiod lacks a nuclear membrane (a defining characteristics of prokaryotic cells) and the DNA is not bound to proteins as it is in eukaryotes. Intact DNA molecules can be prepared for examination in the electron microscope by the Kleinschmidt technique. Autoradiography is another method which allows a direct visualization of the intact bacterial genome. In this procedure DNA is labeled with a radioactive precursor, gently isolated from the cells so as not to break the fragile DNA molecules, collected on a filter, overlaid with a photographic emulsion and developed after proper exposure. Photographic grains represent the contour of DNA. With both these procedures a variety of bacteria have been found to contain only one DNA molecule per cell. Very often, the molecule is in the form of a covalently closed circular ring and replication appears to start at the point of attachment of the ring to the plasma membrane and the two daughter genomes are separated due to a growth of the membrane. *E. coli* cells, when lysed, liberate a circular DNA molecule 20Å wide and about 1200 µ long. This suggests that although the amount of DNA per cell in bacteria is much more than the amount of DNA per virus, its physical organization in both the groups of organisms is the same. In both the cases the genome is represented by a single molecule of the nucleic acid. According to one model the DNA appears to be folded into a number (12 to 80) of super coiled loops. The loops appear to be held in position by a core of RNA, which binds to the DNA and determines the positions of the folds, and the protein. The RNA is newly transcribed single stranded RNA, and the protein is largely RNA polymerase. Nicking with DNase relaxes the supercoiling in one loop without affecting the other loops. Cutting of the RNA molecules in adjacent loops causes the two loops to form one large loop without loss of supercoiling. In addition to the four usual bases, adenine, guanine, thymine and cytosine, bacterial DNA also contains small amounts of methylated bases, e.g. 6-methylaminopurine and 5-methylcytosine.

As a model organism, *Escherichia coli* has 4.6 × 10^6 base pairs in its genome, consisting of a single chromosome or nucleoid. It contains approximataly 4300 genes, compared to humans that have 1,00,000 genes. It is monoploid and generally asexual. The long history of nucleoid isolation from *E.coli* suggests the importance ascribed to successful isolation procedures. Nucleoids prepared by the polylysinespermidine procedure retained the general nucleoid shapes that were present in the cells from which they came and had levels of DNA compaction similar to or greater than those in the cells. such properties support the use of polysinespermidine nucleoids for a number of purposes. The *E.coli* nucleoid maintains its DNA in a compact form yet permits the DNA to carry out vital functions iincluding replication, recombination, gene regulation, and expression. It means, the *E. coli* nucleoid contains DNA in a condensed but functional form. Analysis of proteins released from isolated spermidine nucleoids after treatment with DNase I reveals significant amounts of two proteins not previously detected in wild-type *E. coli*. Partial aminoterminal sequencing has identified them as the products of rdg C and yej K. These proteins are strongly conserved in gram-negative bacteria, suggesting that they have important cellular roles. In another experiment, the DNAase I-treated spermidine nucleoids, which is separated by polyacrylamide gel electrophorasis and identified by either by comigration with known standards or by partial N-terminal amino acid sequencing, reveals five proteins, namely, RNA polymerase, HU, H-NS, Fis and residual lysozyme. Of these, Fis, RNA polymerase, and H-NS are bound particularly strongly to the nucleoids. As previously discussed that there is identified two polypeptides as the protein products of the genes rdgC and yejK. The quantitation of these two polypeptides is based upon the intensity of coomassie blue staining of bands on the membrane and so is subject to some variability due to intrinsic differences in staining and transfer. It is cleared that these two polypeptides are relatively abundant components of the DNase-I-released fraction in all parental and mutant strains tested and represent 2 to 3% of the total protein in this fraction. In the absence of DNase I treatment, there was a greater than 20-folds decrease in the levels of these polypeptides. This strongly suggests that the two polypeptides are associated with nucleoid DNA.

Large Scale Isolation of Plasmid and Bacterial DNA

NP40 LYSIS METHOD FOR CESIUM CHLORIDE GRADIENTS

1. Grow *Escherichia coli* in 500 ml of Luria broth (LB) at 37°C with vigorous shaking. To amplify plasmid DNA, add 2.5 ml chloramphenicol (34 mg/ml in ethanol) when culture reaches an OD600 of 0.40.5 and continue vigorous shaking for 12-16 hrs.
2. Sediment cells at 5520 × g (6000 rpm) in two 250 ml centrifuge bottles for 10 min at 4°C.
3. Resuspend cells in 125 ml (0.25 vol) 10 mM Tris (pH 8.0), 10 mM EDTA and resediment (or wash in 1 M NaCl and resediment).
4. Resuspend cells in 10 ml of 15% sucrose, 50 mM Tris (pH 8.0), 50 mM EDTA, 1 mg/ml lysozyme (freshly prepared), and transfer to a 40 ml Oak Ridge tube. Incubate at room temperature for 10-60 min. Add RNase to 20 μg/ml. Let stand 10 min.

5. Add 10 ml 0.1% NP40, 50 mM Tris (pH 8.0), 50 mM EDTA. Do not vortex. Mix gently by hand, continuously. If lysis does not occur, incubate at 37°C for 30 min.
6. Centrifuge at 35000 × g (17000 rpm) for 50-60 min at 4°C.
7. Decant supernatant into sterile tube until viscous material above the pellet is reached. Around 18 ml supernatant will be obtained. Add 3.7 g of solid CsCl to four different 10 ml sterile tubes. Add 4 ml supernatant to each tube and mix well by gently inverting tube several times. When mixed, add 0.4 ml ethidium bromide (10 mg/ml) to each tube. Density should be about 1.59 g/ml. Each suspension should be sufficient to fill one 5.1 ml rotor tube. Load rotor tube with a needle syringe. Heat seal rotor tubes. Make sure tubes opposite each other in rotor are balanced.
8. Centrifuge at 55000 rpm in rotor for about 10 hr at 20°C.
9. View bands by illuminating tube by longwavelength uv light. Covalently closed circular DNA is in the lower band. Evenly cut off nipple on top of tube, puncture bottom of tube with needle and titrate bands out, collecting the DNA band in their respective tubes.
10. Extract out ethidium bromide from DNA sample with equal volume of 5 M NaCl, TE saturated isopropanol. Remove top isopropanol phase and discard. Repeat extraction until solution is clear in colour.
11. Add 2 volumes water and 6 volumes 95% ethanol to extracted aqueous phase, then place at 20°C for 3 hours or overnight. Centrifuge at 14600 × g (11000 rpm) for 30 min at 4°C, decant supernatant, wash with cold 70% ethanol, recentrifuge, decant supernatant, vacuum dry pellet, and resuspend in 100 µl TE.

Note: This procedure does not work too well for isolation of low copy number plasmids in

Xanthomonas. It may work if volumes are reduced to fill one rotor tube and plasmid amplification is used.

The rdgC gene encodes a slightly acidic polypeptide of 303 amino acids. This protein had not been obseved in wild-type *E. coli*, however a plasmid containing the rdgC gene studied in another study. That study identified rdgC as a gene required for the proper replication of DNA in cells deficient in the recombination enzymes RecABC and SbcCD. The rdgC protein has significant homologs in nine diverse gram-negative bacteria, with levels of amino acid identity ranging from 35 to 92% and levels of overall similarity ranging from 48 to 95%. However, no homologous proteins were found in the complete genomes of four archaebacteria (*Archaeoglobus fulgidus, Methanococcus jannaschii, Pyrococcus horikoshii, and Methanobacterium thermoautotrophicum*) or in the complete genome of the gram-positive bacterium *Bacillus subtilis*. The level of sequence conservation of yejK proteins argues that it plays a significant role in a shared cellular process, its isolation from spermidine nucleoids suggests that this function may invole interactions with nucleic acids. Nucleoids in polylysinespermidine lysates may be useful in gene localization studies, such as those using flourescence *in situ* hybridization or green fluorescent protein derivatives. The regular geometrical shapes of the released nucleoids, particularly those from chloramphenicol-treated cells, suggest their use as standard particles.

Plasmids

Bacteria besides carrying a single circular chromosome formed by a single DNA molecule, they can also carry extra chromosomal material called plasmids, which is also a single DNA molecule in circular form, but only 1-2% the size of the bacterial chromosome. Plasmids are rings of DNA found floating freely in bacterial cytoplasm, outside the main chromosome. The ring of DNA is coiled in a right-handed superhelical coil. There is one superhelical turn for every 400-600 base pairs. The twisted conformation is known as convalently closed circular (ccc) DNA. Cleavage or nicking of one of the two DNA strands results in the release of the twists of the ccc from to give rise to an open circular form. Bacterial plasmids can be transmitted from one cell to another, either in the course of sex. This ability to trade genes with all comers makes bacteria amazingly adaptible; beneficial genes, like those for antibiotic resistance, may be spread very rapidly through bacterial populations. It also makes bacteria favourites of molecular biologists and genetic engineers, new genes can be inserted into bacteria with ease. Plasmids contain very few genes, usually three to four. Plasmids carry genes which determine a variety of biological functions. They may also carry genes which specify the ability to transfer plasmids from donor to recipient cell by conjugation. The essential functions of the cell are coded by chromosomal genes. The bactorial cell does not require any naturally occurring plasmids for viability or growth. Plasmid functions are additional to the functions of chromosomal genes and confer selection advantage to the bacterial cell under varied environmental conditions. Some of them control antibiotic resistance (RTF or resistance transfer factors) in bacteria. Besides the sex and resistance transfer factors, bacteria may also contain colicinogenic factors. Colicinogenic cells (col$^+$) release colicin particles which are capable of killing susceptible strains, col$^+$ cells themselves are immune to death by colicins. Colicin particles can be transferred from col$^+$ cells to col$^-$ cells by conjugation or transduction. Colicin particles have been shown to be supernumerary DNA particles which are also a part of bacterial plasmids. They are infectious and mutiply independently of the main bacterial genome, but under the control of latter as is evident from the fact that only a limited number of replicates (usually two) are permitted and that mutations in the host genome can affect plasmid replication. Although, cytoplasmic, some of them have the capacity to get integrated into the bacterial genome. Those plasmids which can get integrated stably within the bacterial genome are known as episomes. A plasmid may be conjugative or non-conjugative. A conjugative plasmid carries genes that promote the transfer of the plasmid from the host cell to a recipient cell by conjugation. It may also promote the transfer of chromosomal and/or non-conjugative plasmid DNA through a process called mobilization. A non-conjugative plasmid is unable to promote its own transfer by conjugation. Recipient cells which acquire plasmids during conjugation are termed transconjugants. The molecular weight of conjugative plasmids is an excess of 20×10^6, while non-conjugative plasmids arre much smaller. They are apparently not large enough to carry genes coding for plasmid transfer during conjugation.

Most of the plasmids used as vectors. Plasmids containing genes providing immunity to specific colicins or drug rasistance are desirable to permit selection of transformed cells from non-transformed cells. Commonly used plasmid vehicles contain 3,000-30,000 kilobase pairs (kb) of nucleotides.

The first plasmid used or molecular cloning was pSC 101: pstands for plasmid, SC for the initials of Stanley Cohen in whose laboratory it was first used in 1973 and 101 is the strain number. Col E1 is another plasmid which has been extensively used as a vector. It is a plasmid which codes for a bacterial protein, colicin E1, that kills other bacteria which lack it . Many derivatives of these plasmids are also in use these days. Both pSC101 and Col E1 can exist as separate plasmids in the same cell. A large part of this DNA, except the regions which contain replication functions (rep), restriction targets and genes used for selection, can be dispensed with. Plasmids replicate much faster than the main genome and under optimal conditions as many as 1000 to 3000 copies of a plasmid can be found in each cell.

There are at teast 2 types of plasmids: **F or fertility plasmids**, which carries genes for conjugation or DNA transfer. **R resistant plasmid**, which carries genes conferring resistance to antibiotics such as ampicillin, chloranphenicol etc.. AmpR=ampicillin resistant, AmpS=ampicillin sensitive. When plasmids are employed as cloning vehicles or vectors in genetic engineering, they are as following types:

1. **Indicator Plasmids:** If a plasmid lacks a selective feature its cloning can be brought about by co-transformation plasmid. The second plasmid known as Indicator plasmid should carry a gene permitting selection, and should be temperature sensitive with reference to replication. The selective

characteristic of the indicator plasmid is used for selecting the transformed clone. The indicator plasmid can then be eliminated by growth at a restrictive temperature.

2. **Stringent and Relaxed Plasmids:** Plasmid SC101 and its derivatives are called Stringent plasmids, while Col E 1 and its derivatives are called relaxed plasmids with respect to DNA replication. Stringent and relaxed plasmids are 6-8 and 20-30 in number perchromosome respectively in the exponentially growing cells. Plasmid replication requires new protein synthesis in stringent while in relaxed plasmid it is not. In relaxed plasmid, active DNA palymerase I require for replication while in stringent it is not required.

3. **Non-self-transmissible and Self-transmissible Plasmids**: The plasmids pass from one cell to another during conjugation only in the presence of another plasmid which mobilize's them, are known as non-self-transmissible plasmids. For example pSC101 and Col E1. Self transmissible plasmids can be transmitted to many bacterial types. For example members of the P class of plasmids.

4 **Composite Plasmids:** Before foreign DNA is inserted into a plasmid the latter must undergo cleavage. Cleavage and insertion should not take place in the replicator region, or it will become inactivated, and replication of plasmid will not take place. It is possible to produce composite plasmids by inserting drug resistance factors. such as pML2 or $pML2_1$ along with the required gene. Such plasmids will have two cleavage sites Hind III site and Bam HI or sal I sites Cleavage of the plasmid pSC101 at one site by Eco RI and insertion at this site does not affect replication, but may affect tetracycline resistance. Cleavage at the Hind III site does not affect either function. Cleavage and insertion at the Bom HI or sal I sites affects tetracycline resistance. Cleavage of Col E 1 at the Eco RI site does not affect replication, but prevents

production of E 1 colicin. The plasmid, however, still has immunity to the E 1 colicin.

Plasmid DNA occurs in two forms (i) Co-integrates and (ii) Aggregrates.

(i) **Coi Integrate Plasmids:** They have all the resistance genes in one circular molecule together with the genes required for transfer. For example-R100 (Cm, Sm, Tu, Tc) which are isolated from *Shigella flexneri.*

(ii) **Aggregate Plasmids:** They have various genes located on separate molecules. For example, Type 29 isolated from *Salmonella typhimurium.*

R-Plasmids or R-factor

Plasmids do not generally integrate into the host chromosome via recombination. Thus, acquistion of R plasmids resembles bacterial sex in that "snippets" of DNA from other bacteria are making it into the reference bacterium's cytoplasm, but fails to fully resemble bacterial sex in that subsequent recombination into the host chromosome and replacement of host DNA does not occur. By being a means by which fully novel, evolved genes are brought into novel genetic backgrounds, this mechanism of horizontal plasmid DNA transfer contributes to the ability of bacterial species to rapidly adopt to new niches such as those associated with antibiotic resistance. It was long known that repeated exposure of bacteria to antibiotics results in the development of a resistant strain. Some bacterial cells undergo gene mutation which makes them resistant to the antibiotic. They live and multiply in the presence of the drug, giving rise to a resistant strain. This type of resistance develops gradually, and can be treated by administering antibiotics before a large number of resistant cells are formed.

Another type of resistance was observed with respect to the antibiotic gentamycin. Bacterial strains developed resistance to gentamycin only a few months after its introduction. The gentamycin-resistant strain was to also found to be resistant to three other antibacterial agents, chloremphenicol, ampicillin and sulfonamide. Growth of the strain with any susceptible strain of bacteria conferred antibiotic resistance to the latter within a few hours. Thus the antibiotic resistance was infectious. The infectious antibiotic resistance factor or R-factor consists of genes producing resistance to the four antibiotics and genes that facilitate the transfer of the R-factor. They are carried in plasmids which are known as the R (for resistance) plasmids. When a strain carrying R plasmids conjugates with other bacterial cells, a copy of the R plasmid is transferred to the latter, giving it the property of drug resistance.

The control of replication and expression of the cloned DNA is one of the most important aspects of beneficial utilisation of the recombinant - DNA technology. It is in this context that the regulatory regions of bacterial operons form a vital part of the vector DNA. For example, in the modified bacterial plasmid pBGP 123, lac operon genes of *E. coli* and 285 rRNA gene of *Xenopus laevis* are located next to each other. The transcription of *Xenopus* genes is thus regulated by the promoter and operator regions of the bacterial lac operon.

In plants the beneficial utilisation of genetic engineering for the manipulation of nitrogen-fixing (nif) genes is the major concern of those who are striving to maximise crop productivity with least dependence on synthetic nitrogenous fertilisers. The nif genes of *Klebsiella pneumoniae* are expressed in a mutant (when transformed with the recombinant plasmid pMB 9 containing the the nif genes) which lacks the ability to fix atmospheric nitrogen. Similarly, plasmids of *Agrobacterium tumefaciens* can transfer genes directly into the plant cells and can also mediate the transfer of genes from one plant to the other. These genes are expressed in the new environment.

R-plasmids were first discovered in a strain of *Shigella* in the late 1950s during an epidemic of dysentery in Japan. *Shigella* acquired drug resistance with the introduction of antibiotic therapy in 1946. By 1964, 50% of *Shigellas* were resistant to all the four major drugs (Sulfonamides, streptomycin tertracycline and chloromphenicol) against dysentery. Plasmid mediated drug resistance has serious medical implications. Firstly, contact with a drug-resistant non- pathogenic member of the normal gut flora can make a drug-sensitive pathogen resistant. Secondly, a pathogen may acquire new R plasmid resistances during epidemics.

F-factor or Sex Factor

In bacteria, like *E.coli*, sex is determined by the presence or absence of the factor that is called F-factor or sex factor. This was discovered in 1946 by Lederberg and Tatum, about 18 years after transformation was first described by Griffith. The F-factor plasmid is nicked at its origin and replicates as a rolling circle, causing a single-stranded DNA to be produced. It takes a bit of time for the entire F-factor to be replicated, and if the bacteria are interrupted during the act, only the DNA that has made it through the cellular bridge will be transferred. So the ability for the sex pili to form is conferred by genes in a plasmid known as F-factor (fertility factor). Cells with the F-factor are designated $F^+$and cells without it are designated F^-. The first sex factor was recognized in *E. coli* K 12. Sex factor F can either exist as an independent and autonomous unit, or may be intergrated into the chromosome. Because of its dual existence it is sometimes reffered to as the episome to distinguish it from the normal plasmid.

Infectious transfer of bacterial plasmids such as the F factor takes place during conjugation. Donor cells have plasmids in addition to the chromosome. One strand of the plasmid breaks at a specific site and passes into the recipient cell. The single plasmid strands in both donor and recipient cells now synthesize a complmentary strand, giving rise to double- stranded plasmids. Thus identical plasmids are present in both cells, which now become potential donors and are called F cells.

Exchange of DNA segments sometimes take place between the plasmid and the chromosome. Breaks may cccur on both strands of plasmid and chromosome DNA in regions where nucleotide sequences are similar. The two sets of ends join, resulting in the insertion of the plasmid into the chromosome. Cells containing chromosomal DNA plus plasmid DNA rings are called Hfr (for high frequency of recombination) cells. The plasmid may somtimes break away from the chromosome. The breakage and reunion may occur at a point different from the area of incorporation of the plasmid. Under such conditions interchange of segments of DNA takes place between the plasmid and the chromosome.

Genes replicated first from the F factor origin (i.e. those on the 3' side of the origin of replication) are more likely to be transferred to the new host because the conjugation only needs to be maintained for a few seconds. On the other hand, genes replicated last on the rolling circle are less likely to be transferred. In this way, it was possible to construct a genetic map of the chromosome, through purposeful interruption of the conjugating bacteria after specific time intervals, and then determining which genes were transferred at high probability. The conjugating bacteria are interrupted during the act (in the Wollman and Jacob experiment) by putting them into a blender and turning it on to "frappe". If the F factor is excised from an Hfr strain abnormally, a new plasmid is generated that may contain novel sequences. We call this an F' factor. An overexpressing allele of the lac repressor (lacI^q) is present on an F' plasmid. A transposable element called Tn10, carrying a tetracycline resistance marker. If Top10F' cells grow on tetracycline containing medium, they will keep the F' plasmid carrying the lacI^q gene. The lacZdeltaM15 is the C-terminal portion of lacZ that, in combination with lacZ-alpha (the N-terminal portion) in trans, can function as a beta-galactosidase enzyme. The lacZdeltaM15 was introduced into the strain by transduction, on a phage species called phi80. A bit about F factors imagine how a cloning vector could be created that was based on an F factor origin of replication. Such engineered F' plasmids known as "BACs" or Bacterial Artificial Chromosomes. BACs are capable of carrying approximately 200 kbp of inserted DNA sequence, and the F factor origin of replication maintains their level at approximately one copy per cell.

Therefore, the F factor can replicate as an autonomous unit or as an integrated part of the bacterial chromosome. Plasmids with this dual replication ability are known as episomes. Cells containing the F factor integrated into their bacterial chromosome are known as Hfr or high frequency recombination bacteria. These are able to conjugate 1000x times more frequently than F+. Hfr cells can also donate parts of the bacterial chromosome in addition to parts of the F factor, resulting in transfer of chromosomal segments by conjugation. Recipient of Hfr does not necessarily become Hfr because conjugation may be disturbed before entire F factor is entirely transmitted, which is necessary to become Hfr. The F factor is nicked in two during conjugation, so the leading part of the chromosome contains one portion of F, whereas the other end has the second portion.

Since transfer of whole chromosome may take close to 2 hrs, most likely conjugation is disrupted before it happens. The orientation in which F is inserted in the chromosome would determine the polarity of the Hfr chromosome. It can integrate anywhere. Nature of the Hfr strains became clear in 1957 by the work of Wollman and Jacob, who investigated the pattern of transmission of Hfr genes into F-cells. They found that it was possible to disrupt conjugation of two different strains by taking a sample of the bacterial mix and blending it in a kitchen blender. They called this procedure **interrupted-mating.**

Control of Gene Expression in Bacteria

Cells have 2 ways of controlling metabolism: Regulating enzyme activity and regulating enzyme synthesis.

1. **Regulating Enzyme Activity:** The end product of an anabolic pathway may turn off its own production by inhibiting activity of an enzyme at the beginning of the pathway. (feedback inhibition).
2. **Regulating Enzyme Synthesis:** Accumulation of products triggers a mechanism that inhibits mRNA production by genes that code for an enzyme at the beginning of the pathway. (gene repression).

Operons and Their Control

Francois Jacob and Jacques Mond proposed a mechanism for gene regulation, the operon concept.

Structural Gene: gene that codes for a polypeptide.

Operon: a regulated cluster of regulated structural genes with related functions.

(a) common in bacteria and phages

(b) has a single promotor region, so an RNA polymerase will transcribe all structural genes on an all-or-none basis.

(c) contains a single operator, a DNA segment between the operon's promotor and structural genes. It is a binding site for the operon's repressor protein. Acts as the on/off switch for movement of RNA polymerase and transcription of the structural genes.

(d) **Repressor:** specific protein that binds to an operator and blocks transcription of the operon. It blocks the attachment of RNA polymerase to the promotor.

(e) some operons are switched on by activators. DNA-binding regulatory protein which activates transcription of the operon.

(f) **Regulatory Genes:** genes that code for repressor and activator proteins.

Repressible Operon: trp Operon.

If there is no tryptophan in the cell, the regulatory gene produces a repressor protein which remains inactive,allowing the operator to proceed,by allowing the RNA polymerase to complete the formation of the m-RNA and the issuing polypeptides making tryptophan.

If there is tryp in the cell, the repressor protein binds to the promotor with trp as an allosteric inhibitor. This stops the production of the polypeptide.

An Inducible Operon

Lac Operon: If lactose is absent from a cell the enzymes will not be made. If the sugar enters the cell, the operon will be switched on and produce the enzymes needed to digest the lactose.

Lactose metabolism in *E. coli* is programmed by the lac operon which has three structural genes. lac.Z codes for B-glactosidase which hydrolyzes lactose.

lac. Y codes for permease, a membrane protein that transports lactose into the cell, and lac A codes for transacetylase, an enzyme that has no known action in lactose metabolism. Lactose acts as an inducer to turn on the operon.

	Repressible Enzymes		Inducible Enzymes
1.	Their genes are switched on until metabolite activates the repressor	1.	There genes are switched off specific until a specific metabolite inactivates the repressor.
2.	Generally function in an anabolic pathway.	2.	Functions in catabolic pathways.
3.	Pathway end product switches off its own production by repressing enzyme synthesis.	3.	Enzyme synthesis is switched on by the nutrient the pathway uses.

Gene Discovery in Domain Bacteria

Genes with a wide variety of functions were identified in *G. obscuriglobus* and *Pi. marina*, including those of metabolism and biosynthesis, transport, regulation, translation and DNA replication, consistent with established phenotypic characters for these species. The genes sequenced were predominantly homologous to those in members of other divisions of the Bacteria, but there were also matches with nuclear genomic genes of the domain Eukarya, genes that may have appeared in the planctomycetes via horizontal gene transfer events. Significant among these matches are those with two genes atypical for Bacteria and with significant cell-biology implications—integrin alpha-V and inter-alpha-trypsin inhibitor protein—with homologs in *G. obscuriglobus* and *Pi. marina* respectively.

Gene homologs identified were predominantly similar to genes of Bacteria, but some significant best matches to genes from Eukarya suggest that lateral gene transfer events between domains may have involved this division at some time during its evolution. The planctomycetes (order Planctomycetales) comprise a distinct group of the domain Bacteria that forms a separate kingdom-level division on the basis of 16S rRNA analyses. Planctomycetes are currently represented by only a few cultured and characterized heterotrophic members isolated from aquatic habitats; however, the presence of planctomycete sequences in 16S rRNA clone libraries constructed from environmental samples reveals that these organisms occupy diverse ecological niches including waste-water and waste-treatment bioreactors, marine sediments and organic aggregates, and oxic and anoxic terrestrial habitats. In many of these environments planctomycetes make up a significant proportion of the microbial population, indicating that they may have a significant role in the cycling of organic or inorganic compounds. The recent discovery that the "missing lithotroph" responsible for the anaerobic oxidation of ammonium (anammox process) is an autotrophic planctomycete, and the existence of other planctomycetes with similar activities, highlights the potential importance of these bacteria for the flux of nutrients in the environment and indicates the potentially wide physiological diversity of the division.

Consistent with their phylogenetic distinctiveness, the planctomycetes possess a series of unusual phenotypic characteristics common to members of the division. These include budding reproduction, peptidoglycanless (proteinaceous) cell walls and a complex internal ultrastructure. Most notably, cells of at least three species exhibit eukaryote-like membrane-bounded nuclear regions: the genomic DNA of *Gemmata obscuriglobus* is enclosed by two membranes, whereas that of *Pirellula marina* and *Pi. staleyi* is enveloped by a single membrane . Membranes surrounding the nucleoid are unique to the planctomycetes among members of the domain Bacteria. This feature is only one aspect of a unique type of cell organization shared by all planctomycetes examined so far, involving compartmentalization via intracytoplasmic membranes.

Despite the interesting ecological and cell biological aspects of the planctomycetes, molecular studies on this group have been relatively few. To date, DNA sequencing studies on the planctomycetes have involved genes for small subunit (SSU) and large subunit (LSU) rRNA, and a small number of protein-coding genes, including those for the ß-subunit of ATPase from *Pi. marina*, the DnaK heat-shock protein (HSP70) from *Pirellula* and *Planctomyces* species and the gene rpoN for sigma factor 54 from *Planctomyces limnophilus*. Phylogenetic analyses using these gene sequences have failed to elucidate the evolutionary relationship of the planctomycetes relative to the other divisions of the Bacteria. Some sequence analyses show that the planctomycetes form a sister-group of the chlamydiae, while others place them as a deeply branching division. In more recent analyses using the gene for the conserved protein, elongation factor-Tu, inconsistency in the branch position of the planctomycetes was attributed to long-branch attraction effects .

Sequence tags from *G. obscuriglobus* and *Pi. marina* that represent putative protein-coding genes were identified by comparison of individual clone nucleotide sequence translated in all reading frames against protein-sequence databases using the BLASTX algorithm. Within the Bacteria, best matches did not cluster within any one division, although matches with members of Gram-positive and cyanobacterial divisions were common in the case of *G. obscuriglobus*.

Random sequence tags from the two planctomycete species displayed homology with proteins of diverse function, including metabolic enzymes (30 and 38% of total for *G. obscuriglobus* and *Pi. marina* respectively), and proteins involved with transport (7 and 10%), regulation (9 and 10%) and the central processes of translation (9% and 7%) and DNA replication (16 and 10%).

A large proportion of the metabolism and biosynthesis genes identified in *G. obscuriglobus* and *Pi. marina* were homologous to those coding for enzymes involved in amino-acid and vitamin biosynthesis. For *G. obscuriglobus*, close database matches were found to the enzymes isopropyl malate dehydrates, diaminopimelate epimerase, asparagine synthetase and 2-amino-3-ketobutyrate coenzyme A ligase, which are involved in the biosynthesis of leucine, lysine, asparagine and glycine, respectively. In addition, a homolog of threonine deaminase was identified. This enzyme catalyzes the formation of a-ketobutyrate from threonine, an intermediary step in isoleucine biosynthesis. In *Pi. marina*, a gene putatively coding for the NADH/NADPH-dependent enzyme glutamate synthase was found. This enzyme, a glutamine oxoglutarate aminotransferase (GOGAT), is important in the incorporation of inorganic nitrogen into cell material by conversion of 2-oxoglutarate and L-glutamine to L-glutamate.

Homologs of genes involved in the biosynthesis of vitamins and cofactors were also identified in *Pi. marina*. These include glutamate-1-semialdehyde-2,1-aminomutase, which catalyzes the final step in the conversion of glutamate to 4-aminolevulinate (the precursor of tetrapyrrole synthesis), uroporphyrin-III C-methyltransferase and 1-deoxy-xylulose 5-phosphate reductoisomerase (DXP). Uroporphyrin-III C-methyltransferase is involved in the synthesis of both cobalamin (vitamin B12) and siroheme (a cofactor for sulfite and nitrite reductases), whereas DXP catalyzes the initial reactions in biosynthesis of isoprenoids, which are precursors of some vitamins such as vitamin A.

The large proportion of amino-acid and vitamin biosynthesis genes identified in the two planctomycete species suggests that they may be able to synthesize many of these growth factors *de novo*. This finding is consistent with the ability of planctomycetes to grow in oligotrophic conditions and habitats. Many planctomycetes, including *Pi. marina* can be cultured on minimal media with only a modest or no requirement for the addition of vitamins.

Other sequences homologous to metabolic genes of interest that were identified include heptaprenyl diphosphate synthase from *G. obscuriglobus* and glucosamine fructose-6-phosphate aminotransferase (GlmS) from *Pi. marina*. Heptaprenyl diphosphate synthase has been shown in Bacillus species to catalyze the synthesis of the prenyl side chain of menaquinone-7. GlmS reversibly catalyses the formation of D-glucosamine-6-phosphate and L-glutamate from D-fructose-6-phosphate, using L-glutamine as the ammonia source. This is the initial step in a pathway that produces N-acetyl-D-glucosamine as the final product, a compound that is a major component of the bacterial cell-wall constituent peptidoglycan. Interestingly, peptidoglycan is not a component of the predominantly protein cell walls of planctomycetes, including those of *Pirellula* and *Gemmata* species. Therefore, it is possible that the net reaction catalyzed by GlmS in planctomycetes is the reverse reaction—the conversion of glucosamine-6-phosphate to fructose-6-phosphate (an intermediate compound in the glycolysis pathway). Reactions flowing in this direction would be consistent with the ability of planctomycetes to utilize N-acetylglucosamine as a sole source of carbon and nitrogen. Alternatively, it is possible that N-acetylglucosamine is synthesized but utilized in the formation of glycoprotein or polysaccharide rather than in that of peptidoglycan.

A number of genes homologous to those involved in transport across membranes were sequenced from both *G. obscuriglobus* and *Pi. Marina*. Several transport protein gene homologs were also identified in *Pi. marina*. Of particular interest were two genes putatively involved in the synthesis and transport of capsular and O polysaccharide, RfbB and ABCA protein, suggesting that this organism may be able to produce and secrete extracellular polysaccharides or lipopolysaccharides, in a similar manner to Gram-negative organisms.

Several genes putatively involved in signal transduction and regulation of transcription were identified. Homologs of the osmotic shock response regulator OmpR and the nitrogen regulation sensor kinase NtrB were identified in *G. obscuriglobus*. Thus, in this organism, there is evidence for signal transduction proteins that form 'two-component systems', simple regulatory systems operating through protein phosphorylation cascades that allow adaptation to environmental changes.

In *Pi. marina*, there is evidence for the presence of a chemotaxis signal transduction system. Homologs of two proteins involved in bacterial chemotaxis in *Bacillus subtilis*, CheB and CheC, were identified. CheB has been shown to display methylesterase activity and is involved in modification of the cytoplasmic domains of the chemoreceptor methyl-accepting chemotaxis proteins, while CheC is involved in determining the direction of flagellar rotation.

A number of sequence tags with matches to enzymes involved in DNA replication were identified, including several genes putatively encoding helicases. A gene for a replicative DNA helicase (DnaB), which acts at the replication fork by disrupting the hydrogen bonds between the complementary base pairs, was identified in *G. obscuriglobus*. *Pi. marina* also contains a putative DNA helicase that had its closest match with the Hus2 helicase of the eukaryote *Schizosaccharomyces*.

Several sequence tags putatively coding for proteins central to translation were identified. These included two classes of proteins: ribosomal proteins and aminoacyl-tRNA synthetases. In *G. obscuriglobus*, two homologs of ribosomal protein genes were found, one for ribosomal protein L23 and one for L4. In the assembled ribosome, L23 is located within the A site of the 50S subunit and is one of the few ribosomal proteins that directly binds LSU rRNA. In bacteria, the L4 protein is implicated in both ribosomal peptidyltransferase activity and in some cases, autoregulation of the S10 ribosomal protein operon.

Chapter–5 Special Activities of Bacteria

Introduction

Bacteria play numerous important roles in various environmental chemical cycles as well as mineralization of, especially dead animals. The world would be a far different place in the absence of bacteria. Indeed, arguably, eukaryotes could not survive the loss of all the world's bacteria, though one could readily imagine a world consisting solely of bacteria. Such a world, in fact, would be equivalent to that which existed prior to the rise of the eukaryotic lineage, a span which includes a majority of the time on earth during which life existed. As a whole bacteria exploit a far greater spectrum of special bacterial activities. So, it can be said that despite of their simplicity, bacteria have an enormous range of special features which are described below.

The Gram Reaction

One of the most important cytological features of bacteria is their reaction to a simple procedure called, after its discoverer, the Gram stain. Gram staining process was developed in 1884 by Danish microbiologist, Hans Gram. In

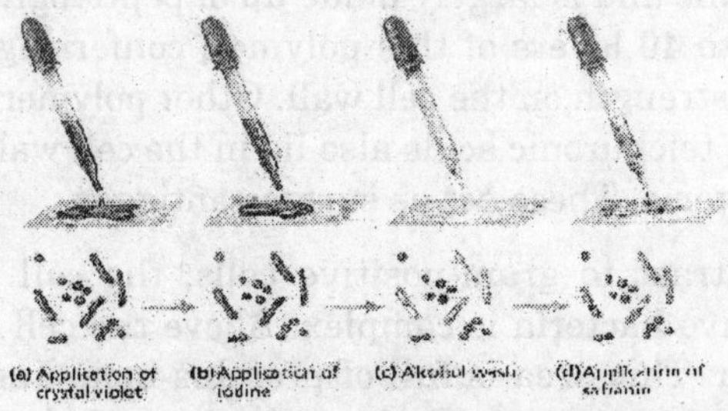

Fig. 15: Gram Reaction Process

this process, the bacteria are stained purple with crystal violet and iodine, rinsed with alcohol to decolorize.

As a result, bacterial cell walls either stain purple or not satined. So, the bacteria which are stained purple known as Gram positive bacteria (Gram+) while those which are not stained known as Gram negative bacteria (Gram-). Gram negative bacteria need counterstaining with another dye to be seen using Gram's method. A red dye such as Safranin or dilute carbon fuchsin are often used. Gram negative bacterial cell wall then stain reddish-pink.

It may seem a very arbitrary basis on which to build eubacteria's classification system. This reaction, however, reveals fundamental differences in the structure of bacteria. Electron microscopy shows that Gram-negative and Gram-positive bacteria have fundamentally different structures, related to the composition of the cell wall, amongst other things. Even eubacteria which give ambiguous gram-stains (or lack cell walls) can often be grouped into gram-negative or gram positive taxa based on other characteristics particularly including DNA sequence data. Gram-positive bacteria have simpler cell envelops than gram-negative bacteria, particularly consisting of a plasma membrane and a thick cell wall. Gram-negative bacteria, in contrast, have cell envelops consisting of a thinner cell wall as well as a membrane external to the cellwall (outer membrane) which is in addition to the plasma membrane found inside the cell wall.

The cell wall of gram-positive bacteria lies beyond the cell membrane and is largely made up of peptidoglycan. There may be up to 40 layers of this polymer, conferring enormous mechanical strength on the cell wall. Other polymers including te choic and teichuronic acids also lie in the cell walls of gram-positive bacteria. These act as surface antigens.

In contrast to gram-positive cells, the cell envelop of gram-negative bacteria is complex. Above the cell membrane is periplasm. This area is full of proteins including enzymes. One or two layers of peptidoglycan lie beyond the periplasm. Gram-negative bacteria are thus mechanically much weaker, than gram-positive cells. Beyond the peptidoglycan of the gram-negative cell wall lies an outer membrane. This has protein channels-poric-through which some molecules may pass easily. The outer side of the gram-negative outer membrane contains lipopolysaccharide. This provides the antigenic structure of the surface of gram-negative bacteria and also acts as "Endotoxin". It is this that is responsible for eliciting the symptoms of gram-negative shock if it gains access to the bloodstream. Porins and Outer Membrane Proteins (OMPs) act as transporters through the outer membrane.

In terms of pathogenicity, gram-negative bacteria tend to be more dangerous than gram-positive bacteria. This is because the body reacts more explosively and badly to gram-negative bacteria, particularly to the above referred "Endotoxin", the component of their outer membrane. The body also has more difficulty clearing gram-negative infections.

A few medically important bacteria do not stain easily using conventional stains, and need to be heated to near boiling point in the chosen dye (carbol fuchsin for light microscopy: rhodamine-auramine for fluorescence microscopy) for at least five minutes. This is to allow the dye to penetrate the waxy cell walls. Having taken the stain, these bacteria resist decolourisation with both acids and alcohol, and are known as acid-alcohol fast bacteria. This is a property of mycobacteria. These include *Mycobacterium tuberculosis*, the cause of tuberculosis; a chronic infection. Most common is pulmonary tuberculosis, affecting the lung. The kidneys may be infected in renal TB, and there is a rare form of osteomyelitis (bone infection) and meningitis caused by TB. In miliary tuberculosis, the infection is disseminated through the body. Another medically important mycobacterium is *Mycobacterium leprae*, the cause of leprosy; a chronic infection of the skin and nerves. Nerve damage leads to a loss of sensation, and ultimately to paralysis. This can lead to tissue damage that can lead to the loss of fingers and toes.

The nature of the cell wall of both bacteria are also reflected in different cell architectures which are shown in following figures.

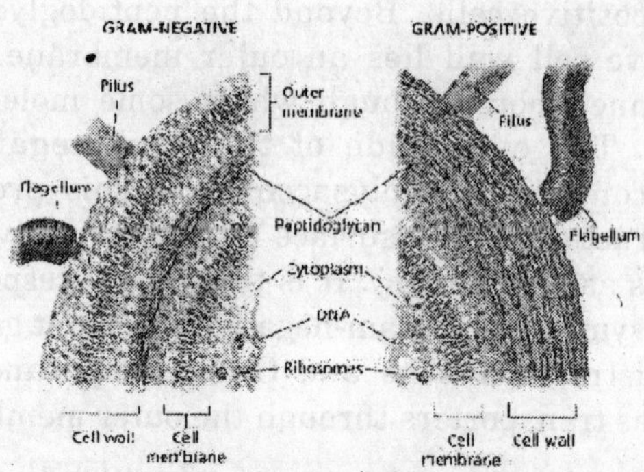

Fig. 16: Comparision between Gram positive and Gram negative Bacteria

The Modes of Motion

Many but not all bacteria exhibit motility, i.e. self-propelled motion, under appropriate circumstances. In bacteria, motion can be achieved by one of following four mechanisms:

I. Flagellar movement;

II. Spirochaetal movement;

III. Gliding movement;

IV. Twitching movement).

Flagellar Movement

Bacterial movement is produced through the action of various surface structures include flagella, pilli, fambriae and glycocaylyx. These surface structures originate outside the cell membranes, sometimes being attached to it, and extend in to the environment.

(i) **Flagella:** Flagella (Latin word for whip), the threadlike projections that enable most types of bacteria to move. Most are 15-20 μm long and are so thin that they can not be visualized with brightfield microscopy. As structural components flagillum has filament, hook, basal body and flagellin protein. The bacteria possessing flagella

have basal bodies embedded in the plasma membrane as anchoring mechanisms and project through the cell envelop into the environment. They are little semi-rigid whips that are free at one end and attached to a cell at the other. The diameter of a flagellum is 20 nm and having a length 10 times the diameter of the cell. Different bacterial species have distinctive arrangements of flagella and unique flagellar antigens. These can be divided into four groups. Polar, or monotrichous, bacteria have a single flagellum that extends from one end of the cell. Amphitrichous bacteria have a flagellum extending from each of the two ends of the cell. Lophotrichous bacteria have several flagella that extend from one or both ends of the cell. Finally, peritrichous bacteria have flagella distributed over the entire bacterial cell. *E. coli* are peritrichous bacteria with typically 6 flagella distributed about the cell.

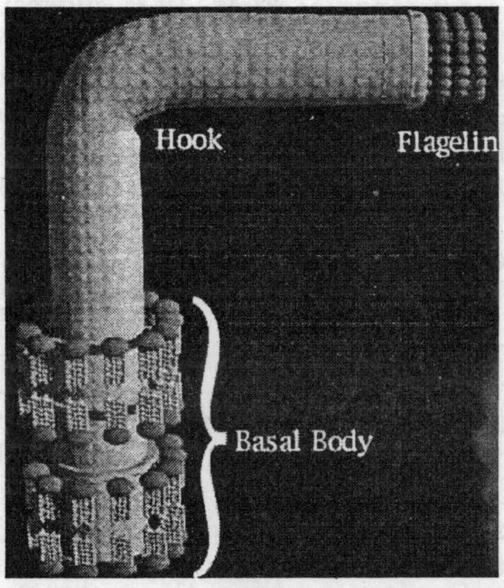

Fig. 17: Flagella of Bacteria

Flagella are mostly composed of flagellin (a protein) that is long chains and wraps around itself in a left handed helix. The number of units, the wavelength and diameter of a single helix of the flagella are determined by the protein subunits. Flagellin is attached to the basal body through the hook which is also protenious structure. The basal body is also called the flagellar motor. The rotation of the flagellar motor has the ability to "switch gears" so that both clockwise and counterclockwise rotation of the flagellum may occur. The force driving rotation is a proton motive force. The exact functionality of this flagellar switch has not been resolved, but it is believed that the chemotaxis protein CheY has a major role in alternating between clockwise and anticlockwise rotation. By default flagellar rotation occurs in the counterclockwise direction. When a sensory signal is received by the bacterium, this signal is transduced by CheY. Phosphorylation of CheA activates it and allows it to bind to the flagellar switch at the base of the motor, causing a switch in rotation directionality.

The flagellar switch consists of three proteins that take part not only in rotation, but also the assembly of the flagellum. These proteins are FliG, Fl:M and Fl:N. CheY binds to the amino terminus of FliM. It is suggested that the possibility of CheY binding to switch the rotational direction is only necessary at room temperature, (Turner, et, al., 1996). The researchers dicovered that at 10°C and below, the motor switches even in the absence of CheY, at near 0°C, clockwise directionality of the motor is actually preferred, as opposed to the default state of counterclockwise rotation that is a characteristic at room temperature.

It is also demonstrated that when the chemoreceptor is unbound (with no attractant), the chemotaxis pathway results in a high level of phosphorylated CheY, which in turn binds to the flagellar motor. This causes clockwise rotation of the flagellum, and a tumbling pattern of motion. When a ligand such as aspartate has bound to the receptor, however, the signal cascade ends with reduced levels of CheY. Thus, the default counterclockwise rotation of the flagellum is allowed to continue, (Melissa Jurica and Barry Stoddard, 1998)

Bacteria move toward attractive stimuli and away from harmful substances and waste products in the process known as Chemotaxis. Chemotaxis is accomplished by sensing the environment and adjusting the rotation of the flagella in response to stimuli. Flagella can rotate clockwise or counterclockwise. When flagella rotate counterclockwise this create a force pushing on the bacteria. In the case of *E.coli* the peritrichous flagella bunch together and all push from one side. This causes the bacteria to move in a straight line, called a run. When flagella rotate clockwise they all pull on the bacteria. With all these forces pulling in different directions, it causes the bacteria to tumble or twiddle. When the twiddling is over, the bacteria will start out a new run in a completely random direction.

Bacterial chemotaxis is controlled by a complex series of events beginning with binding of an attractant molecule to a cell surface chemoreceptor. Chemoreceptor are often clustered at the ends of rod-shaped cells like *E.coli*. Chemoreceptors do not influence flagellarr motion directly, but convey information through a phosphorylation cascade. Information about the environment can be translated into motion within 200 milliseconds. A return to steady-state is assured by a coordinated feedback loop that quickly causes a reversion to original levels of protein phosphorylation in the absence of stimuli.

So, Chemotaxis directs prokaryotic bacterial movement in response to chemical signals in the environment. The purpose for bacterial motility is to give bacteria the capability to continuously move about their media in search of nutrients. Bacteria are attracted to attractants such as sugars (glucose and aspartate) and amino acids. They are repelled by repellents such as toxins and acids (HCl).

In neutral media with no attractants or repellents, bacteria travel through the media in no particular direction, and continue moving in such a random manner until they encounter a chemical gradient. In these circumstances, bacteria tend to move in random directions by rotating flagella in both clockwise and counterclockwise directions. When rotating in a clockwise direction, the flagella spread apart. The force of the many flagella on the cell pulls the cell in many directions, causing a tumbling motion to occur.

When the environment contains an attractant, bacteria sense the concentration gradient of the attractant, and move through the solution/medium in the direction of the attractant. In moving up a concentration gradient toward an area of higher attractant concentration, the peritrichous flagella on the bacteria twist together in order to push the bacteria forward from one side. The bunched flagella rotate as one large whip in a counterclockwise direction. This enables the bacteria to "swim" smoothly and move in a straight line.

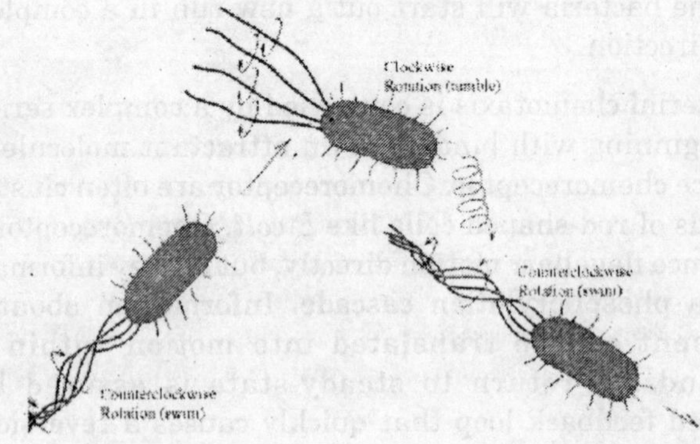

Fig. 18: Illustration of swimming and tumbling

The beating of the flagella creates great force and allows the bacteria to move at speeds of as much as 50um/sec.This is equivalent to 0.0001 miles per hour. For the minuscule size of the bacteria, it is impressively able to propel itself approximately 10 body lengths per second. In relative terms, a cheetah can move at a speed of 25 body lengths per second, while a human athlete can move at about 5.4 body lengths per second. In the presence of a repellent, the bacteria will shorten its "swims" up a gradient, and instead lengthen the "swims" away from the repellent. The response of the bacteria towards a chemical attractant or repellent occurs when a chemical binds to receptors on the cell membrane, which then pass through the cell in a cascade. The discovery of these "swimming" and tumbling patterns have many of their roots in genetic studies. The uses of mutations in examining the chemotactic responses of bacteria have significantly enhanced the scientific understanding of the connection between chemotaxis and flagellar motion.

Mutations in the flagellar motor genes fliG, FliM, and fliN have demonstrated the connection between flagellar rotation directionality and running/tumbling patterns of movement. The behaviour of flagella can be assayed indirectly by attaching the flagella to antiflagellar antibody proteins on glass slides, and observing the motion of the bacterial cell body. Flagellar behaviour is often examined indirectly because of it's small size. Flagella are too thin to observe rotation under the light microscope. Thus, it is convenient to attach flagella to glass slides because flagellar proteins are able to bind tightly to antiflagellar antibody proteins. Once the flagella are bound to the slides, the rotational direction of the cell body can be observed microscopically. Bacterial mutants that phenotypically swim continuously without tumbling tend to rotate in a counterclockwise direction. Mutants that tumble tend to rotate in a clockwise direction. The direction of rotation of the bacterial cell body is the same as the flagella would rotate. Therefore, these experiments provide evidence that counterclockwise rotation of the flagella correlates with "swimming", and clockwise rotation of the flagella results in "tumbling".

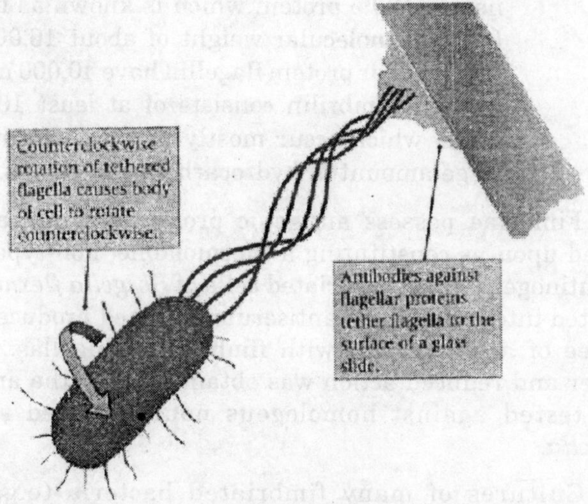

Fig. 19: Flagellar movement of bacteria

(ii) **Fimbriae:** Gram-negative pathogenic bacteria may be covered in fine hairs called fimbriae (singular-fimbria). Fimbriae are shorter and straighter than flagella and are more numerous. All bacteria are not synthesized them. Fimbriae do not function in motility, but are thought to be important in attachment to surfaces or at least in cells sticking together. They can also play a central role in virulence. They may be chemically similar to the flagella. These appendages seem to be restricted to the gram-negative bacteria including *Enterobacteriaceae, Pseudomonadaceae* and *Caulobacter*. In the gram-positive bacteria similar structures have been observed only in *Corynebacterium renale*. The length varies from 0.2-20 µm and their width from 30Å to 140Å. Some have a diameter of 250Å. In *Agrobacterium* the peritrichous fimbriae are 1-3 µm long and 400Å-600Å wide. Fimbriae are also originate from the cytoplasm and penetrate through the peptidoglyan layers of the cell wall. The fimbriae of *E.coli* are nearly 100% protein which is known as fimbrilin. It has a molecular weight of about 16,000, while the flagellar protein flagellin have 40,000 molecular weight. Fimbrilin consists of at least 163 amino acids, which occur mostly in the L Form, with a large amount of hydrocarbon side chains.

Fimbriae possess antigenic properties and have been looked upon as constituting a thermolabile, non-type-specific agglutinogen. When fimbriated cells of *Shigella flexnerii* were injected into rabbits, the antiserum obtained produced a high degree of agglütination with fimbriated shigellas. A much slower and reduced action was obtained when the antiserum was tested against homologous nonfimbriated states of *Shigella*.

Cultures of many fimbriated bacteria (e.g. *E.coli*, *Salmonella typhimurium*) can form a thin layer (pellicle) of cells or star-shaped aggregations on a static liquid medium.

This property is due to the adhesiveness of the fimbriae, and is probably a device for an improved supply of atmospheric oxygen. Fimbriae seem to affect the metabolic activity of bacterial cells. In cultures of *E.coli* K12 the respiratory activity of Fim+ cells is considerably greater than that of Fim$^-$ cells. This supports the view that fimbriae function as aggregation organelles and increase the oxygen supply by forming pellicles on the culture medium. Negative results have, however, been obtained by other workers.

With reference to the presence or absence of fimbriae, at least three types of cells occur in a single colony of *E.coli*.

Fim$^+$: fimbriated cells which are genotypically and phenotypical fimbriate.

Fim($^+$): phenotypically in the nonfimbriate phase but genotypical fimbriate.

Fim$^-$: Mutant strains which remain nonfimbriate under any condition, i.e., which are genotypically as well as phenotypically nonfimbriate.

Conjugation of the Fim- strain with Fim$^+$ or Fim($^+$) can make the former fimbriate.

Ottow (1975) has classified fimbriae into six groups:

Group 1 includes fimbriae which function as adhesive organelles. They range from 100-300 per organism and are peritrichously arranged. Their production is determined by the chromosome. Group 1 fimbriae are divided into four subtypes.

Group 2 includes the sex pili. These are specialized filamentous appendages determined by sex factors or other extrachromosomal episomes. They include the F-pili. Group 2 pili are comparatively scarce, ranging from one to 10 per organism. They are usually longer and wider than fimbriae. They may have a distinct axial canal and may sometimes end in a terminal knob. Some microbiologists put this group in a separate orgon called only pili or sex-pili.

Group 3 includes the thick, hollow tubes of *Agrobacterium*, which may be up to 3 μm long and 400Å-600Å wide. Their function is not understood.

Group 4 includes flexible, rodlike and polarly inserted fimbriae found in *Pseudomonas* and *Vibrio*. Their function is to promote bacterial motion.

Group 5 include the polarly arranged, contractile tubules found in *Pseudomonas 'rhodos; Rhizobium Iupini* and *Agrobacterium* species. They promote conjugation of competent cells by contracting and pulling bacteria together into clusters.

Group 6 fimbriae have been found in the gram positive *Corynebacterium renale*, which has characteristic bundles of fimbriae. Each filament has a diameter of 25-30Å. The filaments act as specific.

(iii) **Pili:** Pili are longer than fimbriae and there are only a few per cell. They are known to be receptors for certain bacterial viruses, but some microbiologists doubt the bacteria make them for that purpose. There are two basic functions for pili, gene transfer and attachment.

Sex pili are filamentous structures which are determined by sex factors (plasmids) such as F, Col I and R factor. These sex factors control the synthesis of the sex pili, their specific

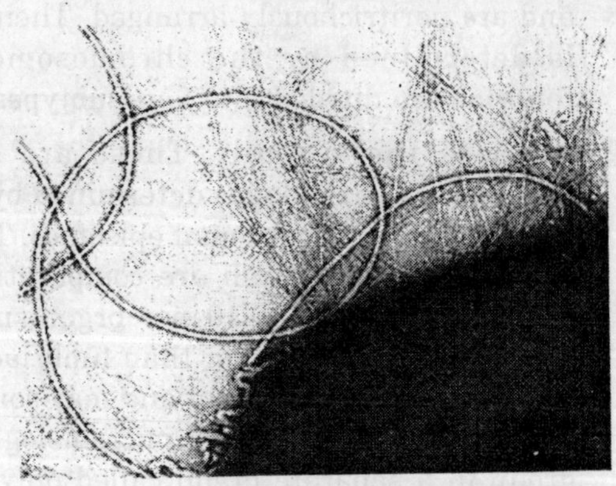

Fig. 20: Transmission electron micrograph of fimbrae and pili

properties, their antigenic features and their capacity to adsorb bacteriophages. The pil have a grater diameter (65-135Å) and length (up to 20μm) than fimbriae, and are usually less numerous (1-10). The sex pilus append to have an axial hole 25-30Å in diameter. Electron micrographs reveal the presence of terminal knobs 150-800Å in diameter. Two types of sex pili can be distinguished in *E.coli,* F pili and I pili. F pili are determined by F or F-like R factors and the I pili by Col I or I-like

R factors. The F pili have receptor sites at which are adsorbed RNA phages such as f2 and MS2 and DNA phages such as f1 and M13. The I pili adsorb the DNA phages If1 and If2, which do not adsorb to the F pili.

The sex pili may provide the means of chromosome transfer during conjugation, probably by acting as conjugation tubes. Immunological studies proved evidence that the tip of the pilus senses the female (F-) cell and constitutes a specific site for attachment. It should be noted, however, that although an axial hole is present in the F-like pili, the I-pili do not appear to have such a hole. Moreover, no DNA has ever been detected in the pilus. The role of the pilus in transfer of DNA is therefore uncertain.

Thus the sex pilus (or F-pilus) is involved in sexual reproduction of certain bacteria. A donor bacteria will attach to a recipient via the sex pilus. Then a copy of part of the donor bacteria's genome passes through the sex pilus into the recipient. This is mechanism of genetic exchange between bacteria. Interestingly, transfer of genes this way is not restricted to species. It is possible for *E. coli* to transfer information to many different gram negative species. Conjugation, as it is called is one explanation for the rapid occurrence of drug resistance in many different species of bacteria.

Pili have also been show to be important for the attachment of some pathogenic species to their host. *Neisseria gonorrheae*, the causative agent of gonorrhea, has a special pili that helps it adhere to the urogenital tract of its host. The microbe is much more virulent when able to synthesize pili.

(c) **Glycocalyx:** Many, but not all bacterial cells have an external coating excreted onto the outside of the cell. There are two types of glycocalyx, capsules and slime layers, but the difference between the two is somewhat arbitrary.

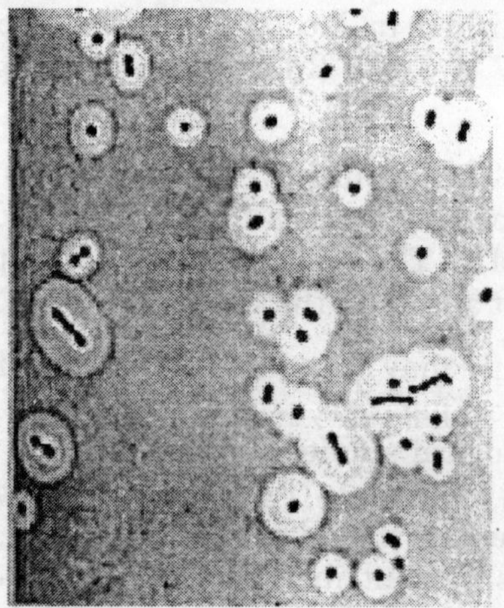

Fig. 21: Capsule surrounding cells of Streptoccus species. The capsule is about the diameter of the cell.

A glycocalyx is a general term for any network of polysaccharide or protein containing material extending outside of the cell. A capsule is closely associated with cells and does not wash off easily. A slime layer is more diffuse and is easily washed away.

Structure

Glycocalyx contains polysaccharide, although some may also contain protein(s), typically glycoproteins. There are many different types of polysaccharides plus some polyalcohols and amino sugars in glycocalyx and the exact makeup is species specific. The structure can be thick or thin, rigid or flexible. Glycocalyx is identified by staining cells with India ink, which does not penetrate the structure. When observed in the microscope the cells appear dark with an outline around them. This outline is the capsule or slime layer.

Function

(a) *Attachment:* These structures are thought to help cells attach to their target environment. Streptococcus mutans produces a slime layer in the presence of sucrose. This results in dental plaque and many bacteria can stick to tooth surfaces and cause decay once S. mutans forms a slime layer. Vibrio cholerae, the cause of cholera, also produces a glycocalyx which helps it attach to the intestinal villi of the host.

(b) *Protection from phagocytic engulfment:* Bacterial pathogens are always in danger of being "eaten" by phagocytes. (Host cells that protect you from invaders.) streptococcus pneumonia, when encapsulated is able to kill 90% of infected animals, when non-encapsulated no animals die. The capsule has been found to protect the bacteria by making it difficult for the phagocyte to engulf the microbe.

(c). *Resistance to drying:* Capsules and slime layers inhibit water from escaping into the environment.

(d) *Reservoir for certain nutrients:* Glycocalyx will bind certain ions and molecules. These can then be made available to the cell.

(e) *Depot for waste products:* Waste products of metabolism are excreted from the cell, and will accumulate in the capsule. This binds them up, and prevents the waste from interfering with cell metabolism.

Spirochaetal Movement: Spirochaetes are helical bacteria which have a specialized internal structure known as the axial filament which is responsible for rotation of the cell in a spiral fation and consequent locomotion. (e.g. *Rhodospirillum*). The axial filoment is a periplasmic structure situated in the space between the inner and the outer membranes of the cell envelop. Axil filaments are similar to flagella in structure and composition. They also have basal body like structure embeded in the inner membrane from this axil filament is produced and attached with a hook region. The molecular weight of the filamental protein is 37,000 which has cyclic aminoacids present in low quantities.

It has been postulated that the protoplasmic cylinder may rotate on the body axis in the opposite direction. This rotation may be due to the rotation of axil filaments in the periplasmic space. The external cell wall layer may rotate in a direction opposite to that of the protoplasmic cylinder which is semi rigid. Due to this semirigidity it permits rotation on the local body axis. It is assumed that due to the propagation of a surface wave bacteria may do free swimming. Similarly it is said that due to the rotation of the outer cellwall layer bacteria may do surface creeping. These all types of the movements are included under spirochaetal movement of bacteria.

Gliding Movement: Gliding motility is the movement of cells over surfaces without the aid of flagella, a trait common to many bacteria, yet the mechanism of gliding motility is unknown. The gliding movement is happened only on the solid moist surface. It is observed that gliding movement is common in gram negative bacteria like Sulfur and Iron-oxidizing bacteria (ex. *Beggiatoa*), Gliding *Myxobacteria* (ex. *Myxococcus*), Cyanobacteria (ex. *Oscillatoria*). Bacteroides/ Flavobacteria (ex. *Flavobacterium*) and Cytophaga (ex. *Cytophaga, Sporocytophaga* and *Flexibacter*). Gliding movement is also found in cell wall less, gram positive bacteria like *Mycoplasma* and *Spiroplasma*. Excreted slime (microfibrils), probably internal contractile fibrils, and other similar protrusions on the cell surface, are supposed to give this type of motility. Doetsch and Hageage have grouped the hypotheses for gliding mechanisms into four categories:

(a) Gradient of osmotic forces along the cell;

(b) Generation of surface tension due to localized secretion of surface active material;

(c) Pushing of the cells by localized secretion of slime;

(d) The contractile wave hypothesis, according to which there is generation of contractile waves in the bacterial cell;

The contractile Wave Hypothesis is best supported on the basis of available evidence.

Twitching Movement: A number of bacterial species including some species of *Pseudomonas* and particularly the grama-negative bacteria *Moraxella* and *Acinetobacter* possess twitching motility. Twitching motility is a flagellum-independent surface translocation which is held on solid media. This surface bound motion is due to the presence and function of thin, usually polar, fimbriae on the cell surface. Such bacteria usually produce spreading colonies which of ten form channels on the agar medium.

The Spore Formation

The spore formation is usually triggered by detrimental environmental growth conditions usually limitation of carbon-supply, starvation for certain nutrients dehydration conditions or accumulation of toxic wastes. The formation of a spore is an expensive and complex process for the bacterial cell. All bacteria cannot form spores. But several types that live in the soil can form the spores. Gram-positive bacteria in the *Bacillus* and *Clostridium* groups are spore formers. Their spores are called endospores. Basically endospores are so called dormant stages of bacterial cells. Different bacterial genera have evolved different dormant structures: these include:

(a) Endospores as in *Bacili* and *Clostridea;*

(b) Exospores (conidiospores) as in Actinomycetes;

(c) Cysts as in Azotobacter;

(d) Non-culturable (but viable) forms as in Enterobacteriaceae. These soil dwelling bacterial groups face very often prolonged periods of scarce or absent C-supply and have therefore evolved a mechanism to survive these suboptimal living conditions. In these circumstances they respond by entering into a different differentiation status called dormant stage in which cell division comes to a halt and all metabolic activities are minimized until the environmental conditions become more favourable.

Endospore formation is not a reproductive method, since one bacterial cell yields one spore, which in turn, yields one bacterial spore, hence it is a survival technique. So, it can also say that endospores are dormant alternate life forms. Excluding *Bacillus* and *Clostridium,* several other less common genera of bacteria also produce endospores, these are *Desulfotomaculum, Sporocarcina, Sporolactobacillus, Oscillospira* and *Thermoactinomyces*.

As it is clear that an endospore is not a reproductive structure but rather a resistant, dormont survival form of the organism. Endospores are quite resistant to high temperatures (greater than 120°C for several hours), radiation (UV and gamma), chemical treatment (EtoH), most disinfectants, dying etc. Endospores are produced within cells ard are refractile-light cannot penetrate them so that they are very easy to see in the phase microscope. They are resting structures, meaning that there is little or no metabolism inside the spore and it is a real form of suspended animation. Endospores can survive for a very long timc, and then regerminate. Endospores that were dormant for thousands of years in the great tomes of the Egyptian pharohs were able to germinate and grow when placed in a appropriate medium. There are even claims of spores that are over 250 million years old being able to germinate when placed in appropriate medium. These results have yet to be validated.

Endospore Structure: The endospore structure can be enhanced selective spore stains and by more detailed structures are revealed by studies under the electron microscope. The completed endospore consists of multiple layers of resistant coats surrounding a nucleoid, some rebosomes, RNA molecules, and enzymes. There are main three parts of an endospore (1) core (2) cortex and (3) coats.

1. **Core:** The core is dehydrated cytoplasm containing DNA, Ribosomes, enzymes etc. Diiplocolinic acid (DPC) is only found in large quantities in the spore core but is not found in vegetative cells. This acid is found in association with Ca which is a chelater. Calcium-dipicolinic acid is deposited in the spore core and represents approx. 10 % of the dry weight of the endospore. Besides its abundance in calcium-dipicolinic acid, the spore core has only 10-30% of

the water content of the vegetative cell. Bacterial endospores derive much of their longevity and resistance properties from the relative dehydration of their cores. Spores are resistant to heat, radiation, chemicals and dessication due to the dehydration of the protoplast and the production of special proteins that protect the spores DNA.

The internal pH of dormant spores (pH 7) of *Bacillus* species is more than 1 pH unit below that of vegetatively growing cells (pH 8) but rises to that of growing cells in the first minutes of spore germination. Acidification of the core during endospore formation seems to be important in the accumulation of storage compounds. Mature endospores are highly resistant to UV-radiation and exhibit unique DNA photochemistry. UV-radiation of spore DNA results in formation of spore photoproduct (SP), the thymine dimer 5—thyminyl—5, 6—dihydro thymine. Repair of SP occurs during spore germination by two distinct routes: either by the general nucleotide excision repair (uvr) pathway or by a novel SP-specific monomerization reaction mediated by the enzyme SP lyase. The SP repair enzymes are synthesized during endospore formation and are packaged in the dormant spore.

The DNA in dormant spores of *Bacillus* species is associated with alpha/beta-type small, acid soluble proteins (SASP), which are double-stranded DNA- binding proteins

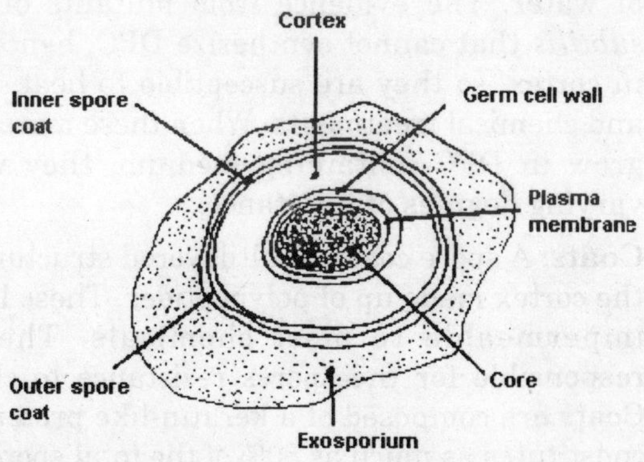

Fig. 22: Detailed structure of Bacterial spore

whose amino acid sequence has been highly conserved in evolution. Binding of alpha/beta-type SASP causes DNA to assume an A-like conformation. (Alpha/beta-type SASP are responsible for packaging the chromosome of the developing spore into a doughnut like structure). Associated with the conformational change in DNA brought about by alpha / beta-type SASP binding is a change in its photochemistry such that ultraviolet irradiation does not generate pyrimidine dimers, but rather spore photoproduct.

2. **Cortex:** The cortex is modified cell wall/ peptidoglycan layer that is not as cross-linked as in a vegetative cell. The spore cortex is responsible for maintaining core dehydration. The cortex may osmotically remove water from the interior of the endospore and the dehydration that results is thought to be very important in the endospore's resistance to heat and radiation. Peptidoglycans in the cortex are similar to those in the vegetative cell walls except for two differences. (1) The spore cortex has a side-chain of L-alanine in every second sub unit of Glucosamine (2) Diaminopimelic acid in tetrapeptide side-chains are only found in the spore cortex. Dipicolinic acid (DPC) is also found in the cortex layer of the endospores. There is a complexing of peptidoglycans with DPC and calcium at the later stages of sporulation. This is accompanied by a loss of water. The evidence from mutants of *Bacillus subtilis* that cannot synthesize DPC, hence no DPC in cortex, so they are susceptible to heat, radiation and chemical treatments. When these mutant spores grow in DPC-containing medium, they will have varying degrees of resistance.

3. **Coats:** A spore coat is multilayered structure around the cortex made up of polypeptides. These layers are impermeable to most chemicals. The coat is responsible for the spores resistance to chemicals. Coats are composed of a keratin-like proteins which constitutes as much as 80% of the total spore protein. The spore coat proteins are rich in cysteine and

hydrophobic amino acids. Sometimes on outer membrane composed of lipid and protein and called an exosporium is also seen. This thin exosporium around the spore coat is a characteristic feature of *Bacillus cereus* group of bacteria.

Sporulation: The process of endospore formation is known as sporulation. In *Bacillus subtilis*, sporulation is a developmental process under genetic control. The decision to either grow vegetatively or sporulate is regulated by the state of phosphorylation of the Spo OA transcription factor. SpoOA is a member of the response regulator family of two-component regulatory systems. Unlike most family members, SpoOA does not directly obtain its phosphate from a histidine protein kinase. Rather, atleast three histidine protein kinases transfer phosphate to the relay protein, SpoOF, then to SpoOB and finally to SpoOA. In addition, the phosphorylation state of SpoOA is modulated by specific phosphatases, such as SpoOE, which dephosphorylates SpoOA-P, and RapA, which dephosphorylates SpoOF-P. SpoOA-P is both a repressor and an activator of transcription depending on the promoter it is affecting.

The phosphorelay is the major regulator of sporulation initiation in *Bacillus subtilis.* A myriad of signals, both positive and negative, from the environment, cell cycle and metabolism is received and interpretated by the phosphorelay and integrated through the opposing activity of protein kinases and protein aspartate phosphatases to create an extremely sophisticated regulatory network. Because commitment to sporulation has serious cellular programming consequences and is not undertaken capriciously, the phosphorelay is subject to a variety of complex controls. Induction of sporulation leads to the derepression of many genes encoding proteins necessary for spore formation. Studies into the molecular events of sporulation are still in their infancy although many genes have now been cloned. However the entire process is very complex and may eventually be useful for studies in cell differentiation in higher eukaryotes. The biochemical trigger for the onset of sporulation is not clear. It must involve de-repression of spore-forming genes and repression of genes that activate the major metabolic pathways. There are 26/27 known genes that are involved in this process. Once genes are activated/repressed, a number of biochemical steps occur. Low metabolism of carbon and nitrogen-source utilization. This is followed by phosphorylation of unknown protein which can lower the levels of ATP. It has also been suggested that sporulation is triggered by cyclic GMP. Cyclic GMP is produced in large quantities during the beginning of sporulation. Cyclic GMP could act like cyclic AMP (analogue) that decreases enzyme synthesis in gram-negative bacteria. This implies that cGMP inhibits enzyme synthesis, leading to lowered metabolism. Regulation of sporulation is tight and the first few steps are reversible. This helps the cell conserve energy and only sporulate when necessary.

Steps of Sporulation Process: Sporulation is a seven step process, although Stage I really doesn't seem to be genetically important.

1. **Stage 0 (Normal Growth):** The first stage of sporulation is involved in forming a separate compartment for the spore in the mother cell. Once this ocuers, sporulation is irreversible.
2. **Stage I (Bacterial Division):** In this stage firstly the DNA replicates and a cytoplasmic membrane septum forms at one end of the cell.
3. **Stage II (Assymetric Separation):** This unequal cell division results the formation of a fore spore septum. So the two progeny cells different in size are produced. The smaller cell is called the forespore and the larger cell is called the mother cell. The forespore and the mother cell do not separate but remain side by side.
4. **Stage III (Engulfment):** In a phago cytosis-like process the forespore becomes engulted by the mother cell. Eventually the forespore is fully surrounded by the mother cell (therefore the name: endospore).
5. **Stage IV (Cortex synthesis):** As it is cleared in the above previous stages that two layers first cytoplasmic

membrane septum at one end of the cell and second forespore septum are formed. Both of these membrane layers then synthesize peptidoglycan in the space between them to form the first protective coat that is cortex. Both the spore and the mother cell play a role in this process.

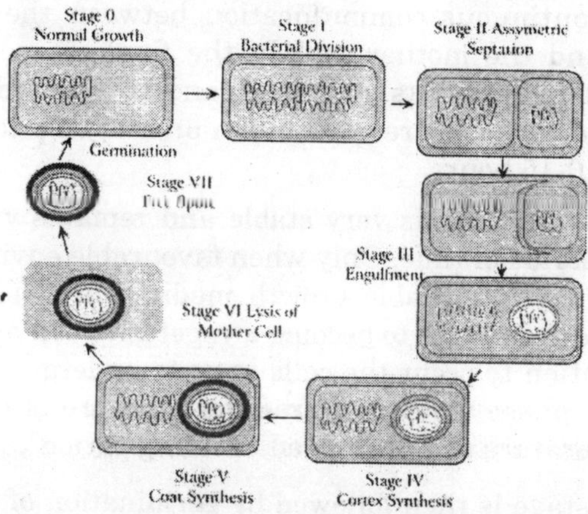

Fig. 23: Stages of spore formation in Bacteria

6. **Stage V (Coat synthesis):** After forming the cortex, the next stage involves laying down a layer composed of a keratin like protein around the cortex layer. This layer known as spore coat.

7. **Stage VI (Lysis of Mother Cell):** When spore maturation is complete (after about 6-8 hours of development) the fully matured spore is released into the environment by lysis of the mother cell. Thus, the mother cell is mortal in that it undergoes "programmed cell death", whereas the forespore is immortal in that it becomes the spore and gives rise to vegetatively growing cells.

8. **Stage VII (Free Spore):** The conversion of the forespore into a mature spore is mediated by events occurring both within the forespore and in the mother cell. Thus, protein products produced within the forespore (such as the family of SASPs)

contribute to the spore core. Conversely, a complex set of proteins that are produced within the mother cell is deposited around the outside of the mother cell membrane that envelops the forespore to create the proteinaceous coat, which will encase the spore. Coordination of these events is achieved by continuous communication between the forespore and the mother cell. In the final stage, the spore dehydrates its cytoplasm and is released from the cell. The entire sporulation process only takes about 10-15 hours.

The spore itself is very stable and remains viable for a very long period of time. Only when favourable environmental conditions and a suitable growth medium become available will a spore germinate to become a vegetative cell again. Prior to germination to occur the cells have to undergo a socalled " activation" phase which requires the exposure of the cells to high temperature or a prolonged " resting period".

This stage is then followed by germination of the spores which is initiated by water uptake and swelling. The biochemistry of the cell alters completely leading to increased respiration and increase in enzymatic activities. At the same time the high termperature resistance is lost. The last stage involves the outgrowth of the vegetative cell leaving the spore coat behind. The spore germination process completes the cycle.

Another group of bacteria called methylosinus produces spores called exospores. The difference between endospores and exospores is mainly how they form. Endospores form inside the original bacterial cell as described above. Exospores form outside by growing or budding out from one end of the cell. Exospores also don't have all the same building blocks as endospores, but they are similarly durable.

Members of *Azotobacter, Bdellovibrio, Myxococcus* and *Cyanobacteria* groups form protective structures called cysts. Cysts are thick-walled structures that, like spores, protect bacteria from harm, but they are somewhat less durable than endospores and exospores. Basically cysts are resting cells which formed from the vegetative cells of bacteria. Like endospores, there is no measurable metabolic activity. Cysts also have large quantities of calcium but no diplocolinic acid (DPC), here calcium confers resistance to desiccation. Cysts show a high degree of resistance to chemical treatments. Unlike endospores, cysts don't show any resistance to high temperatures. Cysts can be considered as vegetative cells that have been modified in three ways:

I. Loss of flagellar motility;

II. Change in shape from elongated to round;

III. Formation of a multilayered coat consisting.

(a) **Intine:** an internal cyst layer made up of lipids and carbohydrates,

(b) **Exine:** an external cyst layer made up of lipoproteins and lipopolysaccharides.

Cyst has a very large granules called polyhydroxybutyrate granules in its centre. These granules are reserve materials that are used up in the switch from aerobic to anaerobic metabolism. Cyst formation is a very slow process, it takes ± 36 hours for completion.

Genetic Recombination

Generally bacteria multiply in a straight forward manner (apologies to molecular biologists for the simplification). Each single celled bacterium grows until there is enough material to form two separate bacteria. The one parent bacterium then splits into two progeny bacteria. This process is known as binary fission. The time that it takes one bacterium to accumulate enough material to split is known as the generation length. This generation length varies greatly between different genus and species of bacteria, from as short as twenty minutes for *E.coli* to as long as twenty four hours for *Mycobacterium tuberculosis*. The population growth curve for bacteria is an exponential curve. With each generation, the number of bacteria doubles. In ideal circumstances, like a ready supply of nutrients and a benign environment, a single *E. coli* bacterium can grow to become over one million bacteria in as little as three and a half hours. For one *Mycobacterium tuberculosis* bacterium to generate the same number of bacteria, again under ideal circumstances, may take as long as ten days. In practice, however, this is not always the case, since the circumstances in which bacteria grow are not always ideal. Nutrients may not be plentiful, the temperature may be too warm or too cold, the acidity of the environment may be too high or too low, there may be chemicals present which inhibit the growth of bacteria or there may be organisms present that consume and destroy bacteria.

Every bacterium has a set of genes that completely describe the bacterium, and which dictate the physical, chemical and biological characteristics of the bacterium. These genes are made of the chemicals DNA and RNA. This set of genes is known as the genotype of the bacterium. Usually, when a parent bacterium splits into two bacteria, the two progeny bacteria are genetically identical.

Though bacteria neither undergo meiosis nor mitosis, and do not require cellular fusion in order to initiate reproduction (indeed, bacteria are not diploid), many bacteria nonetheless maintain active sex lives. This is done through the transfer of snippets of genomic DNA from cell to cell by various mechanisms. Following uptake, these DNA srippets may be incorporated in to the recipient cells DNA by a process of genetic recombination. In other words, the basic mechanism of sex, recombination between DNA sourced from different parents, are employed by bacteria.

Rcombination involves the cutting and covelent joining of DNA sequences. Recombination can occur between two different DNA molecules (Intermolecular Recombination) or between two regions of a single DNA molecule (Intramolecular Recombination). Intermolecular recombination can occur with a variety of types of DNA templates: the DNA templates may be linear DNA acquired via transduction, transformation, or conjugation, or linear chromosomes and plasmids; in bacteria with circular dsDNA chromosomes or circular plasmids, the recombination may occur between linear and circular dsDNA templates or between two circular dsDNA templates.

Various mechanisms are utilized by which DNA may be taken up by bacteria. These mechanisms range from the active uptake to passive mechanisms which result in random gene exchange. It is not certain just what, if any, short term evolutionary advantage bacteria receive from engaging in sex. However, repair of DNA damage certainly has been suggested.

Recombinants of bacterial cell have identified by their growth characteristics, in which media they are able to grow, which is determined by their genetic make up, not morphologically. Petri dishes containing media are used for growing bacteria. For a typical culture, several million bacterial cells are spread, which will grow in a few hours in a solid cover or lawn across the dish. To look at the genetic makeup of individual cell, by serial dilutions one plates a maximum of 300 cells per plate. Each bacterial cell is able to produce a colony by clonal reproduction. The members of colony having a single genetic ancestor are known as clones. Since transfer one colony at a time would be extremely tedious, a simple technique for transferring all the colonies in one plate to other plates simultaneously was devised by bacterial geneticist Joshua Lederberg. This process is known as replica plating. It consists in using a sterile piece of velvet pressed down gently to a dish with the colonies, then use the velvet as a rubber stamp to transfer the colonies to fresh dishes. The original colonies can be followed up by their position and arrangement in the culture dishes. In order to assess colony growth, there is need to characterize individual colonies, whether they grow or not in a particular media. Besides minimal medium, other media used by bacterial geneticists are:

(a) **Complete Medium:** It contains not only the minimal media ingredients, but also supplements not required by prototrophs but needed by auxotrophs.

(b) **Minimal + Specific Supplements:** If the mutant is unable to synthesize proline, then this amino acid can added as a supplement.

(c) **Lactose Plate:** Changes the source of carbon, instead of glucose which is normally used, lactose is used instead. These plates test for the ability of bacteria to utilize lactose as energy source. Those unable to use lactose as C source are called Lac^-.

Some examples are shown in the cartoons below. The double-stranded DNA (dsDNA) is represented by a single line, with the donor dsDNA and the recipient dsDNA shown in black, recombinant DNA shown in black line. An X represents areciprocal cross-over between the two dsDNA molecules and . a half-X represents a nonreciprocal cross-over between the two dsDNA molecules.

Intermolecular Recombination

- A single, reciprocal cross-over can occur between two linear dsDNA molecules as shown in panel A below, or a single, nonreciprocal cross-over can occur between two linear dsDNA molecules as shown in panel B below.

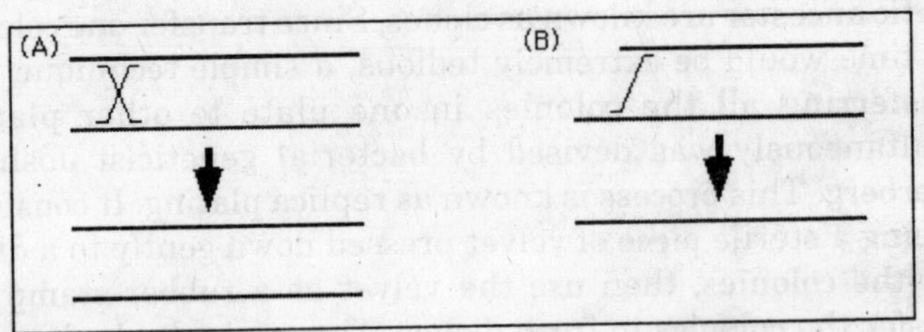

Fig. 24: Single reciprocal and non-reciprocal cross over

- A double, reciprocal cross-over can occur between two linear dsDNA molecules as shown in panel C below, or a double, nonreciprocal cross-over can occur between two linear dsDNA molecules as shown in panel D below.

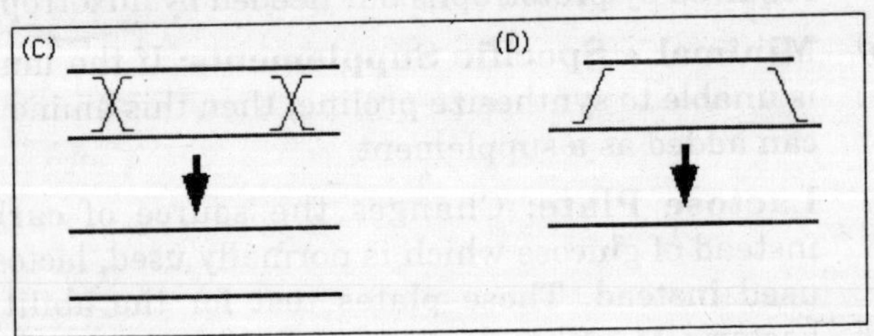

Fig. 25: Double reciprocal and non-reciprocal cross overs

A single, reciprocal cross-over between a linear dsDNA molecule and a circular dsDNA molecule results in a linear dsDNA molecule. In most bacteria with a circular chromosome, such a recombination event would be lethal because the chromosome could not replicate properly and it would be ultimately degraded by DNA exonucleases. Likewise, recombination between a linear DNA molecule and a circular plasmid would destroy the plasmid.

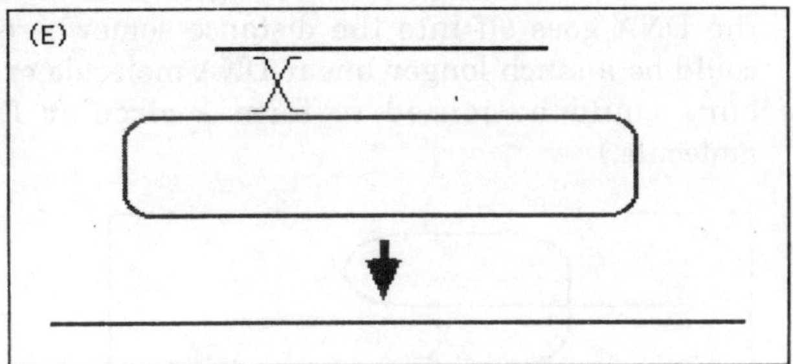

Fig. 26: A single reciprocal cross-over between linear dsDNA and a circular dsDNA

Two reciprocal cross-overs can occur between a linear dsDNA molecule and a circular dsDNA molecule as shown in panel F below, or two nonreciprocal cross-over can occur between two linear dsDNA molecule and a circular dsDNA molecule as shown in panel G below.

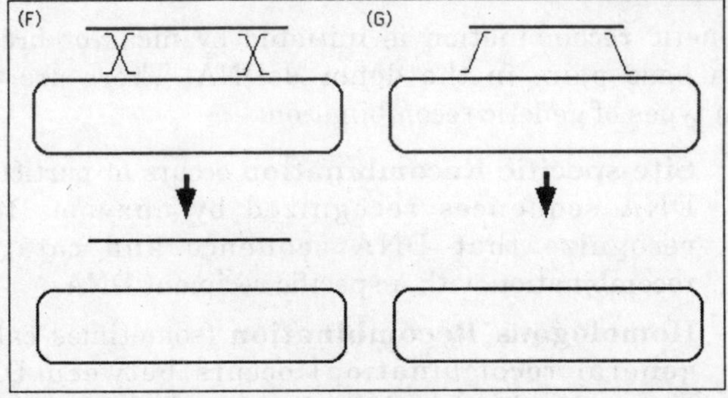

Fig. 27: Two reciprocal and non-reciprocal cross-overs

Intramolecular Recombination

- Recombination between two regions on a single dsDNA molecule can result in looping out of the intermediate region, yielding a shorter dsDNA molecule and a separate circular dsDNA molecule. Note that the reverse of this reaction can also occur, resulting in the integration of a circular dsDNA molecule into another dsDNA molecule. (The slashes at the ends of the lines in this figure indicate that the DNA goes off into the distance somewhere—it could be a much longer linear DNA molecule or the ends could be joined to form a circular DNA molecule.)

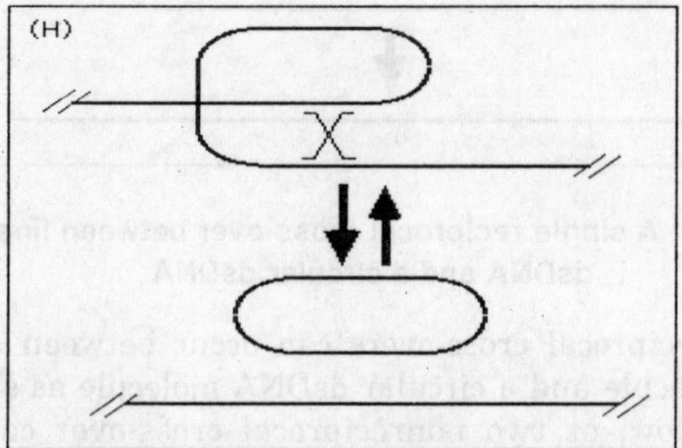

Fig. 28: Intramolecular recombination

Other Types of Recombination

Genetic recombination is initiated by nicks or breaks between base pairs in the donor dsDNA. There are two common types of genetic recombination:

1. **Site-specific Recombination** occurs at particular DNA sequences recognized by enzymes that recognize that DNA sequence and catalyze recombination with a specific recipient DNA.

2. **Homologous Recombination** (sometimes called general recombination) occurs between DNA sequences that are homologous—that is, the two DNA sequences are nearly identical.

The following discussion refers to homologous recombination.

Recombination Frequency

Although there are some sites that are hot spots or cold spots for homologous recombination, when you consider long stretches of dsDNA the probability of a recombination event occurring between any two base pairs in the dsDNA is relatively uniform. That is, in a stretch of dsDNA the breaking and joining reaction could occur between any set of base pairs, as indicated in the figure 29.

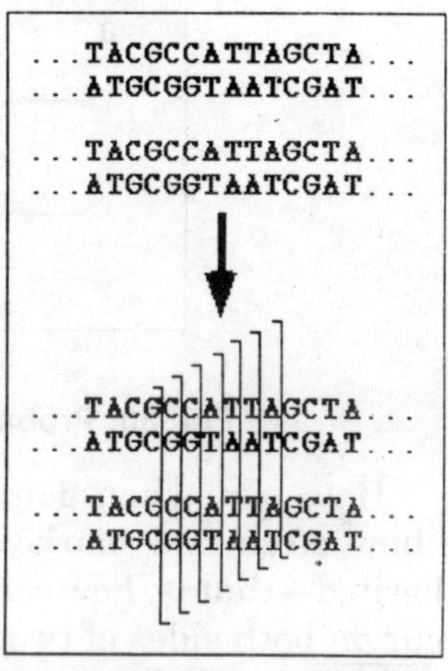

Fig. 29: Set of basepairs

Given an equal probability of recombination between any 2 base pairs, a longer region of homologous DNA will provide many more opportunities for a recombination event. Hence, recombination between two DNA positions that are very close to each other occurs less often than recombination between two DNA positions that are farther apart—or the further apart two DNA positions are the greater the probability that there will be a recombination event between them. (If the two DNA molecules are exactly identical, it would not be able to determine that recombination occurred by genetic approaches because the substrates and products would have the same genotype. Differences in DNA sequences that can be distinguished are called "markers".) The take-home point is that recombination frequencies are directly proportional to distance between two markers. Note that the recombination frequencies are not simply additive. In the figure below there is a 10% probability that a recombination event will occur somewhere between the two markers A and B, and a 2% probability that a recombination event will occur somewhere between the markers B and C, but a 25% probability that a recombination event will occur somewhere between the markers A and C.

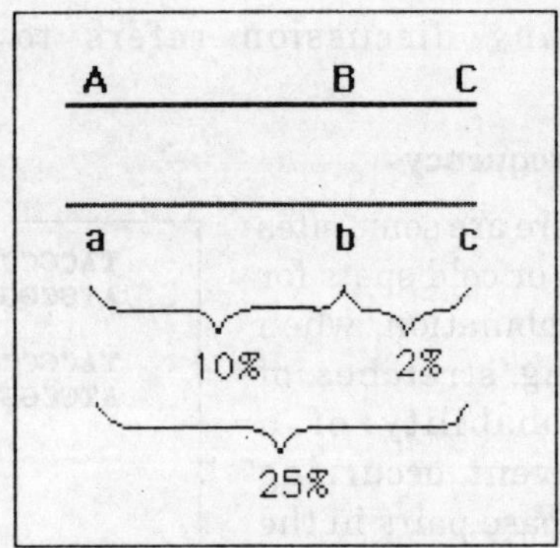

Fig. 30: Probability for recombination

However, when doing genetic crosses we typically look at how often two markers on one dsDNA molecule are co-inherited—that is, how common are recombination events that occur on both sides of two markers so that both markers from the donor are incorporated into the recipient DNA molecule. For example, in the following figure, a recombination event with a cross-over at (i) and at (ii) would result in inheritance of A+ but not B+ or C+, a cross-over at (i) and at (iii) would result in co-inheritance of A+ and B+ but not C+, and a cross-over at (i) and at (iv) would result in co-inheritance of A+, B+, and C+. Co-inheritance frequency is inversely proportional to the distance between two DNA markers. This is a very important concept that will be used extensively when we discuss genetic mapping. Because A is closer to B than to C, co-inheritance of A and B will occur much more frequently that co-inheritance of A and C.

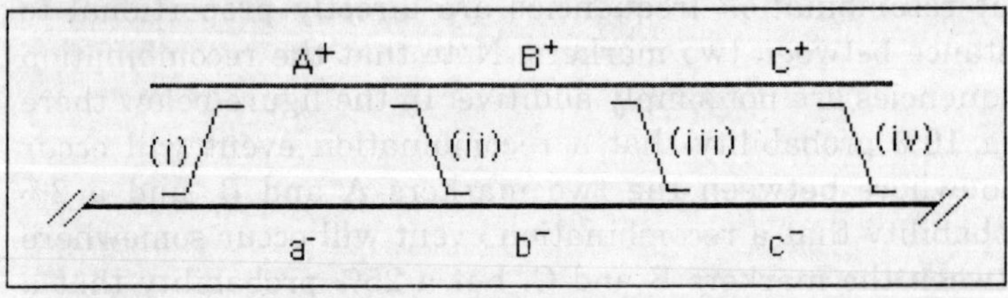

Fig. 31: Coinheritance frequency

Enzymology: Many different enzymes are required for genetic recombination. The following list includes some of the important players in homologous recombination in enteric bacteria.

- RecBCD = The RecBCD complex initiates recombination between two homologous dsDNA molecules when one of the molecules has a free, double-stranded end. The RecBCD complex binds to the free end and rapidly degrades the DNA. When RecBCD encounters a short DNA sequence called a chi site, the RecD subunit is ejected and the RecBC complex now acts as a helicase, unwinding the donor DNA. The resulting ssDNA can invade a recipient dsDNA molecule with the help of RecA protein. (There are other enzymes in enteric bacteria that can catalyze similar functions, so recBC mutants are not completely defective for recombination.)
- RecA—The RecA protein promotes pairing between a single-stranded DNA molecule and a homologous sequence in another DNA molecule, a process called synapsis. (There are no other proteins in enteric bacteria that can catalyze this reaction, so recA mutants are "completely" defective for recombination.)
- RuvAB and RecG—These enzymes catalyze branch migration, resulting in pairing of two homologous strands of DNA.
- RuvC and RusA—These enzymes are endonucleases that cut the "cross-over" (Holliday junction), completing the recombination process.

1. Homologous recombination requires DNA sequence homology between the donor and recipient DNA.
2. Recombination beteween any two sequences is a relatively low frequency event.
3. The co-inheritance of two genetic markers is inversely proportional to the distance between them.
4. Recombination is a complex process involving many different enzymes, and many redundant functions.
5. Homologous recombination requires RecA protein.

By participating in sex, bacteria may be capable of more rapid evolutionary leaps than they would otherwise be capable of partcularly, novel mutational combinations may be achieved through bacterial sex. Though operating less frequently per individual than that achieved among eucaryotes participating in obligate sexual reproduction, the much shorter generation times and high numbers associated with bacteria can nevertheless make such transfers significant contributors to the evolution of bacterial populations. It is almost certain that DNA exchange between related bacteria can maintain the cohesiveness of bacterial species. In other words, some form of the biological species concept may actually operate among otherwise a sexually reproducing bacteria. Exchange of DNA also occurs when plasmids are transferred by various processes from bacteria to bacteria. Exchange of plasmid DNA does not necessarily result in recombination and therefore in increased species cohesiveness. By being a means by which fully novel evolved genes are brought into novel genetic backgrounds, this mechanism of horizontal plasmid DNA transfer contributes to the ability of bacterial species to rapidly adapt to new niches such as those associated with antibiotic resistance.

Sex among bacteria (less strictly, i.e., gene exchange with or without recombination) therefore appears to contribute to the ability of bacteria to adapt to changing environments, genetically explore unoccupied niches such as that associated with antibiotic resistance in pathogenic bacteria and help maintain the cohesive of bacterial species:

There are three ways by which bacteria can obtain DNA from other bacteria:

Conjugation: Direct contact and transfer of DNA from one bacteria to another. A donor bacteria transfers DNA to a recipient through a tube (pili).

Transformation: Direct absorption of free DNA living bacteria, most likely from lysed bacteria, resulting in stable genetic change by recombination. The recipient bacteria resulting from this process are called transformants:

Trransduction: Involves a 2nd organism or shuttle organism such as a phage, a virus able to infect bacteria as a parasite.

The transfer of DNA is mediated by a phage, there is no direct contact between two bacterial cells. There are two basic types of bacterial cells in *E. coli* F is a factor enabling bacteria to donate its double stranded circular DNA plasmid. The F factor initiates a transfer tube or pilus (pili, the plasmid is opened up, and as it is replicated, as it is pushed through the tube into the recipient cell, independently of the bacterial chromosome. This process is known as rolling circle replication and takes 15 to 20 minutes. At the end of the conjugation process, a copy of the F factor remains in the donor, and a copy is transferred to the recipient. Sometimes, the F factor gets incorporated into the bacterial chromosome by crossing over, due to the presence of common homologous regions called insertion sequences (IS).

Conjugation: It was first discovered in 1946 by Lederberg and Tatum in *Escherichia coli.* It has also been observed to occur in strains of other bacteria, such as *Salmonella, Pseudomonas, Shigella* and *Vibrio*. After cell contact, actual chromosome transfer occurs; the process is unique to bacterial conjugation and is vastly different from chromosome transfer in higher organisms. Sexual conjugation in bacteria involves the union of cells designated as male and female. Electron micrographs of paired cells show them connected by a short tube, 50 nm wide and it has been assumed that this pairing represents conjugation. It has also been suggested that the fimbriae attached to the surface of male cells are responsible for the transfer of DNA from the male to the female cells. It has been suggested that a sex factor is attached to the plasmalemma at the point where a fimbria forms.

Recombination frequencies from F^+ X F^- crosses were never particularly high. In the early 1950s, some *Escherichia coli* strains were isolated in which the frequency of recombination was upto 10,000 times greater than observed in F^+ strains. These were called Hfr strains. The discovery of Hfr strains made the fact of the actual physical act of conjugation feasible and a number of fine electron micrographs showing a conjugation bridge between bacterial cells have been obtained.

The discovery of Hfr strains also enabled Jacob and Wollman (1962) to determine the mechanism of transmission of the bacterial chromosome. When a female cell receives a sex factor either by pairing or by fimbrial conjugation, it changes into a male cell and subsequently can conjugate with another uninfected female cell. During the conjugation process, the male cell slowly injects a genophore into the female cell in such a way that the genes enter the female cell in a linear way. The episome becomes attached to the female genophore and is the last part to enter the receptor cell. It takes about 100 minutes for a complete transfer of the genophore to occur. The transfer is usually not complete so that the resulting zygote is usually a partial diploid. Of course, crossing over between bacterial chromosome and exogenous DNA will then occur, thus resulting in recombination.

Conjugation can only occur between cells of opposite mating types.

The donor (or "male") carries a fertility factor (F^+).

The recipient ("female") does not (F^-).

F is a set of genes originally acquired from a plasmid and now integrated into the bacterial chromosome. It establishes the origin of replication for the chromosome. A portion of F is the "locomotive" that pulls the chromosome into the recipient cell. The rest of it is the "caboose". In E. coli, about one gene gets across each second that the cells remain together. (So, it takes about 100 min for the entire

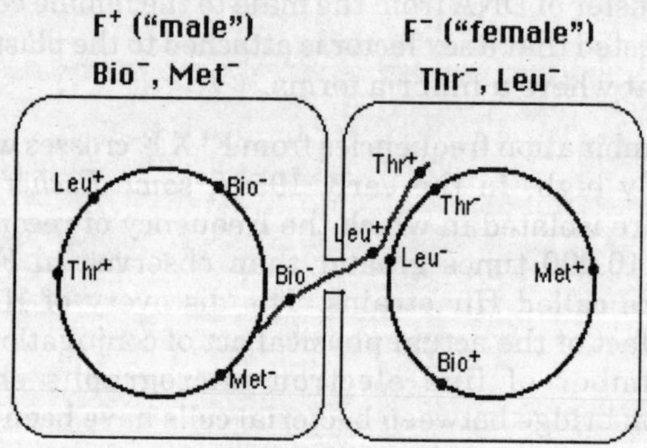

Fig. 32 : Mechanism of conjugation in *E. coli.*

genome (4377) genes to make it. But, the process is easily interrupted so it is more likely that host genes close behind the leading F genes ("locomotive") will make it than those farther back. The "caboose" seldom makes it so failing to receive a complete F factor, the recipient cell continues to be "female". The DNA that makes it across finds the homologous region on the female chromosome and replaces it. By deliberately separating the cells (in a kitchen blender) at different times, the order and relative spacing of the genes can be determined. In this way, a genetic map—equivalent to the genetic maps of eukaryotes—can be made. But here the map intervals are seconds, not centiMorgans.

The above figure shows the mechanism of conjugation in E. coli cells where

- The "male" lacks functional genes needed to synthesize vitamin biotin and the amino acid methionine (Bio-, Met-) so these must be added to its culture medium.
- The "female" has those genes (Bio+, Met+) but has nonfunctional (mutant) genes that prevents it from being able to synthesize the amino acids threonine and leucine (so Thr-, Leu-) these must be added to its culture medium).
- When cultured together, some female cells receive the functional Thr and Leu genes from the male donor.
- A double crossover enables them to replace the nonfunctional alleles.
- N The cells now can grow on a "minimal" medium containing only glucose and salts.

In most bacteria there are several pieces of DNA. One is the somatic genome—a huge circle of double-stranded DNA that actually measures about 2 mm in length, and is all crammed into the little cell. This large piece of DNA is what defines the type of bacterium it is. The cell cannot live without this circle of DNA. In addition, there are various optional smaller circles of DNA, which are usually called plasmids.

A rod-shaped bacterium such as E. coli. Within its envelope consisting of two membranes with an intervening cell wall, are the chromosome and plasmids, if any.

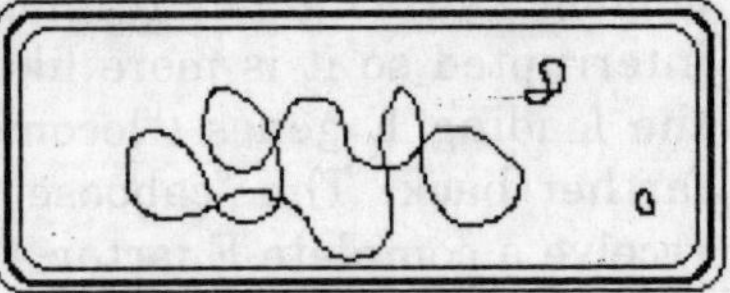

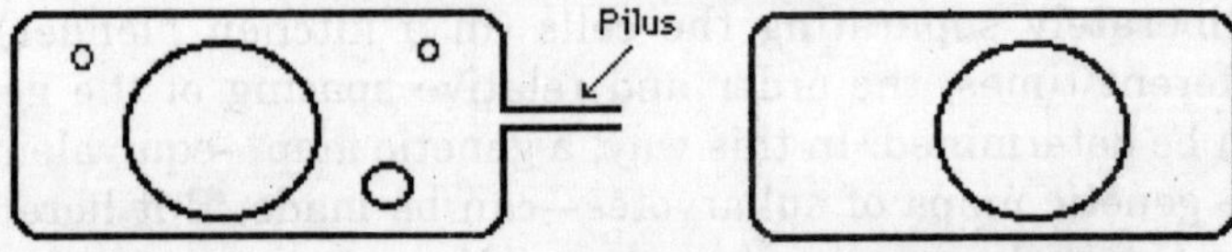

Fig. 33: Donar cell **Fig. 34: Recipient cell**

Above are shown two types of E.coli cells. The DNA's have been schematically untangled and stretched out into circles. The cell on the left has a small plasmid which has coded for a pilus. The cell on the right has no plasmid and no pilus; this cell is a potential recipient, while the left cell is a donor, which can duplicate the plasmid and then send a single-stranded DNA copy over to another cell via the thin tubular pilus.

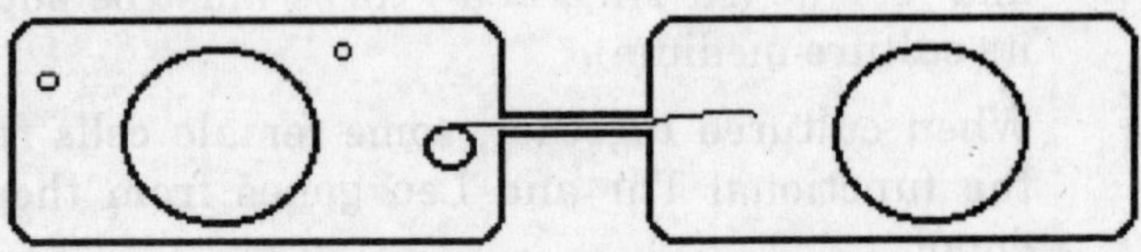

Fig. 35: Conjugation through pilus

Seemingly, each type of plasmid codes for its own special type of pilus. An E.coli cell can have up to several hundred pili, and some plasmids code for the cell's having more than one for its transmission.

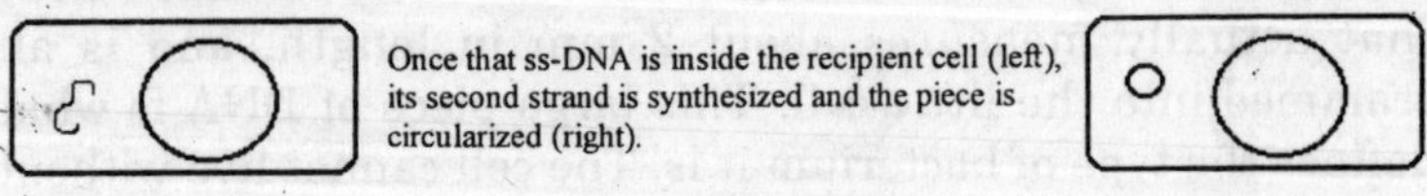

Fig. 36: Conjugant cells

A widely used plasmid of E. coli is one called "F" (for fertility). .. Cells that possess "F" are called male ("F^+").

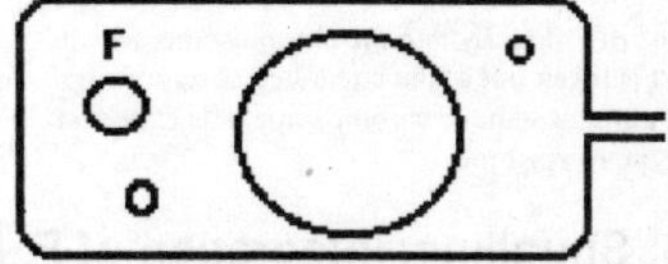

Fig. 37: F^+ cell

Shown here is an F^+ cell that also has two other plasmids in it, which we shall soon disregard.

These cells usually possess two F-pili for the transport of the F into cells lacking it. The cells shown here only have one for simplicity of drawing. Those without F are called females (potential recipients, F^-). Like any self-respecting plasmid, the F also promotes its transmission to receptor cells, converting them to being F^+ within a minute or two.

As Joshua Lederberg noted (and for which he was awarded the Nobel Prize), one of the characterists of F and a few other plasmids is that it can carry additional genes—such as for antibiotic resistance—and often a number of those genes all in a row. Since the plasmid can so quickly be spread from one cell to another, it is little wonder that antibiotic resistance is building up so rapidly in the world's bacteria.

But let us focus on F's ability to transfer a more benign gene such as the lac-operon, which confers upon the bacterial cell the ability to metabolize the dissaccharide lactose. Thus if F also includes the lac-operon; its donation to a recipient that previously could not use lactose allows the recipient quickly to begin using that sugar.

Interestingly, it can be easily shown that a "good" lac-operon can undergo recombination with an inoperable one in the main chromosome.This means that the cell contains two alleles of the lac-operon: one that works and the other which is defective - or, one that is dominant, and one that is recessive.

F lives two lives—one is as a free-living plasmid floating around in the cytoplasm of the cell. But the other life is one of even more profound implications to life on earth.

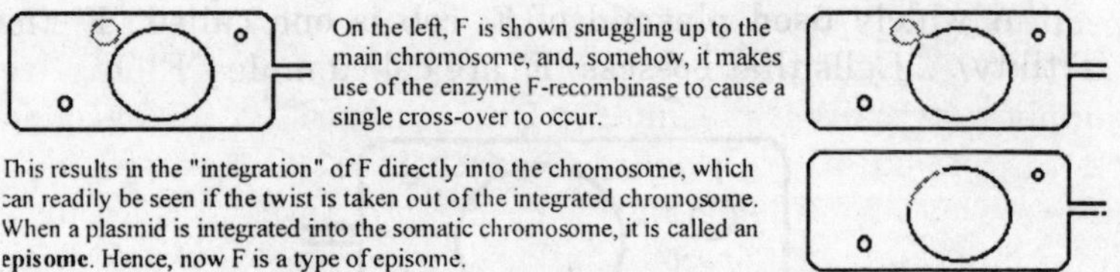

Fig. 38: Snuglling, Integration of F, Episome

Conjugation is the unidirectional transfer of genetic information between cells by cell-to-cell contact. The term "unidirectional" refers to the fact that a copy of the plasmid is transferred from one cell, termed the "donor", to another cell, termed the "recipient". There are two dissimilar functions of plasmids involved in conjugative ability: the first is a site of initiation of transfer that is called either oriT or mob. The oriT is short for "origin of transfer" and mob is for "mobility". In each case they refer to a site on the DNA and not to a diffusible product. The second group of functions are those proteins that act at these sites and cause the range of functions necessary for mobilization to occur. These are encoded by the tragenes and have a variety of functions including the formation of the pilus that makes contact with the recipient cell and seems to be involved in drawing the donor and recipient cells together. This brings about a region of membrane contact and it appears a conjugation bridge of some sort is formed. The products of the tra genes are involved in both the regulation and the physical construction of these events. Some event in this sequence triggers the nicking of a site, termed oriT, by a specific single-strand nuclease and a subsequent binding of one or more "pilot" proteins to the free 5' end of the DNA. These proteins seem to function in the subsequent replication of the transferred DNA with one serving as a "primase". A single-strand is then transferred from this end to the recipient while a "rolling circle" form of replication occurs in the donor. If the DNA being transferred is a plasmid, it is made double-stranded and circularized in the recipient, whereupon it can presumably replicate. If the transfer DNA is chromosomal, circularization cannot occur, but in some way a complementary strand is generated and homologous recombination with the chromosome can occur (in any case, the incoming DNA must become associated with a replicon if it is to be inherited). It is possible for a plasmid to be non-conjugative and yet mobilizable (if the traproducts are supplied by another plasmid) so long as an oriT site is encoded on the plasmid (whoever has the oriT site is transferred). Finally, a plasmid lacking both the trafunctions and oriT functions would be non-conjugative and non-mobilizable. Many applications in inverse genetics employ small plasmids containing oriT regions, where the tra functions are supplied by another plasmid in the cell. Plasmids have mechanisms that increase the likelihood that, following cell division, both daughter cells will contain a copy of the plasmid. The partition functions (often termed par) responsible for this work by a variety of mechanisms including monomerization of plasmid multimers (better to have many monomers than a few multimers) and association of the plasmid with membranes (which apparently helps physically separate the plasmids). While we refer to a plasmid being "lost" by a cell, the actual mechanism is almost certainly that the cell never received the plasmid at the previous cell division due to inappropriate partitioning. Such loss is termed segregation. For both low and high copy-number plasmids, this "loss" occurs at (very roughly) 1% frequency, though some exceptionally stable plasmids have been found, presumably because of a set of different par functions. Some plasmids have evolved a system, with effects like the par systems above, that "prevents" segregation by killing any daughter cell that has not received a plasmid. They do this by producing a relatively long-lived killing function (kil) and a short-lived kill override (kor) function. A daughter without the plasmid will have the kil product, but will not be able to maintain the necessary amount of kor product to survive. To the experimenter, these systems will look like partitioning systems, since, in mutants lacking these, plasmid-free segregants will be more frequently detected. They can also appear to be inc functions.

Occasionally, it is necessary to isolate plasmid-free derivative of a strain currently containing a plasmid, a procedure termed curing. These can be sought (i) spontaneously (perhaps replica printing isolated colonies if the plasmid confers a scorable phenotype); (ii) following an enrichment (again, if the plasmid confers a growth phenotype); (iii) by selecting a different, but incompatible, plasmid into the cell; (iv) or by treatment with elevated temperature or chemicals such as acridines, ethidium bromide, sodium dodecyl sulfate and novobiocin (since the first two chemicals are known as mutagens, they should be used with restraint).

Some bacteria, *E. coli* is an example, can transfer a portion of their chromosome to a recipient with which they are in direct contact. As the donor replicates its chromosome, the copy is injected into the recipient. At any time that the donor and recipient become separated, the transfer of genes stops. Those genes that successfully made the trip replace their equivalents in the recipient's chromosome.

Transformation: "Transformation" means change. In genetics, transformation refers to a process in which bacteria take up free DNA and incorporate it into their genomes. Bacterial transformation was discovered in 1928 by an English bacteriologist, Fred Griffith, who was studying the polysaccharide capsule of *Streptococcus pneumoniae*. This bacterium causes pneumonia, but in order to do so, it must be able to produce a capsule which prevents white blood cells from engulfing and destroying it. The cells of *S. pneumoniae* (also known as the pneumococcus) are usually surrounded by a gummy capsule made of a polysaccharide. When grown on the surface of a solid culture medium, the capsule causes the colonies to have a glistening, smooth appearance. These cells are called "S" cells. Encapsulated strains produce smooth, shiny colonies because of their slimy capsules. Some mutants of *S. pneumoniae* fail to produce either the capsule or disease. These are called "rough" strains or "R" cells, because their colonies appear rough on agar plates. Griffith injected dead, smooth *S. pneumoniae* and live, rough *S. pneumoniae* into mice. The mice died of infection. Control mice injected with either dead, smooth S. pneumoniae or with live, rough S. pneumoniae alone, survived. In the experimental mice, either the live cells brought the dead cells back to life, or the dead cells somehow changed (transformed) the live cells so that they could make capsules.

Streptococcus pneumoniae (pneumococci) growing as colonies on the surface of a culture medium. Left: The presence of a capsule around the bacterial cells gives the colonies a glistening, smooth (S) appearance. Right: Pneumococci lacking capsules have produced these rough (R) colonies.

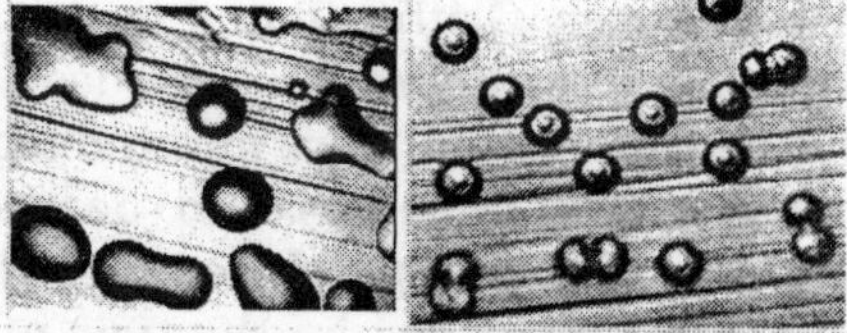

Fig. 39: Streptococcus *pneumoniae*

With the loss of their capsule, the bacteria also lose their virulence. Injection of a single S pneumococcus into a mouse will kill the mouse in 24 hours or so. But an injection of over 100 million (100 × 106) R cells is entirely harmless.

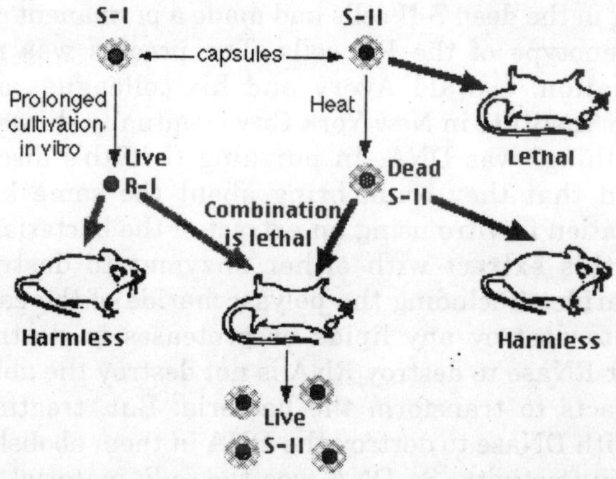

Fig. 40: Diagramatic presentation of experiment on mice

The capsule prevents the pneumococci from being engulfed (by phagocytosis) and destroyed by the scavenging cells—neutrophils and macrophages of the body. The R forms are completely at the mercy of phagocytes.

Pneumococci also occur in over 90 different types: I, II, III and so on. The types differ in the chemistry of their polysaccharide capsule.

Unlike the occasional shift of S -> R, the type of the organism is constant. Mice injected with a few S cells of, say, Type II pneumococci, will soon have their bodies teeming with descendant cells of the same type.

However, Griffith found that when living R cells (which should have been harmless) and dead S cells (which also should have been harmless) were injected together, the mouse became ill and living S cells could be recovered from its body. Furthermore, the type of the cells recovered from the mouse's body was determined by the type of the dead S cells. In the experiment shown, injection of

- living R-I cells and
- dead S-II cells

produced a dying mouse with its body filled with living S-II pneumococci. The S-II cells remained true to their new type. Something in the dead S-II cells had made a permanent change in the phenotype of the R-I cells. The process was named transformation. Oswald Avery and his colleagues at The Rockefeller Institute in New York City eventually showed that the "something" was DNA. In pursuing Griffith's discovery, they found that they could bring about the same kind of transformation in vitro using an extract of the bacterial cells. Treating this extract with either enzymes to destroy all polysaccharides (including the polysaccharide of the capsule) or lipase to destroy any lipids or proteases to destroy all proteins or RNase to destroy RNA is not destroy the ability of their extracts to transform the bacteria. But, treating the extracts with DNase to destroy the DNA in them abolish their transforming activity. So DNA was the only material in the dead cells capable of transforming cells from one type to another. In fact, their experiments provided one of the most convincing proofs that DNA is the genetic material.

Although the chemical composition of the capsule is determined by genes, the relationship is indirect. DNA is transcribed into RNA and RNA is translated into proteins. The phenotype of the pneumococci—the chemical composition of the polysaccharide capsule—is determined by the particular enzymes (proteins) used in polysaccharide synthesis.

Bacterial transformation is the process by which bacterial cells take up naked DNA molecules. If the foreign DNA has an origin of replication recognized by the host cell DNA polymerases, the bacteria will replicate the foreign DNA along with their own DNA. When transformation is coupled with antibiotic selection techniques, bacteria can be induced to uptake certain DNA molecules, and those bacteria can be selected for that incorporation. Bacteria which are able to uptake DNA are called "competent" and are made so by treatment with calcium chloride in the early log phase of growth. The bacterial cell membrane is permeable to chloride ions, but is non-permeable to calcium ions. As the chloride ions enter the cell, water molecules accompany the charged particle. This influx of water causes the cells to swell and is necessary for the uptake of DNA. The exact mechanism of this uptake is unknown. It is known, however, that the calcium chloride treatment be followed by heat. When E. coli are subjected to 42°C heat, a set of genes are expressed which aid the bacteria in surviving at such temperatures. This set of genes is called the heat shock genes. The heat shock step is necessary for the uptake of DNA. At temperatures above 42°C, the bacteria's ability to uptake DNA becomes reduced, and at extreme temperatures the bacteria will die.

The process for the uptake of naked plasmid and bacteriophage DNA is the same; calcium chloride treatment of bacterial cells produces competent cells which will uptake DNA after a heat shock step. However, there is a slight, but important difference in the procedures for transformation of plasmid DNA and bacteriophage M13 DNA. In the plasmid transformation, after the heat shock step intact plasmid DNA molecules replicate in bacterial host cells. To help the bacterial cells recover from the heat shock, the cells are briefly incubated with non-selective growth media. As the cells recover, plasmid genes are expressed, including those that enable the production of daughter plasmids which will segregate with dividing bacterial cells. However, due to the low number of bacterial cells which contain the plasmid and the potential for the plasmid not to propagate itself in all daughter cells, it is necessary to select for bacterial cells which contain the plasmid. This is commonly performed with antibiotic selection. *E. coli* strains such as GM272 are sensitive to common antibiotics such as ampicillin. Plasmids used for the cloning and manipulation of DNA have been engineered to harbor the genes for antibiotic resistance. Thus, if the bacterial transformation is plated onto media containing ampicillin, only bacteria which possess the plasmid DNA will have the ability to metabolize ampicillin and form colonies. In this way, bacterial cells containing plasmid DNA are selected.

The transformation of bacteriophage M13 into bacterial cells is identical to plasmid DNA transformation through the heat shock step. After the heat shock step, single stranded M13 DNA begins replicating in the host cell through use of the host cell machinery. During the life cycle of this virus, however, M13 replicative form is created and daughter phages are packaged and extruded from the bacterial cell. These intact phage molecules then infect neighbouring bacteria in a process called transfection. When these transformed and transfected bacteria are plated with non-infected cells onto growth media, the non-infected cells form a background cell lawn which covers the plate. In regions of M13 transfection, areas of slowed growth, called plaques, can be identified as opaque regions which interrupt the lawn.

Since M13 viral transfection is a critical part of the transformation of bacterial cells with M13, it is absolutely necessary to use a strain of *E. coli* which harbors the episome for the F pilus. When M13 phages infect bacterial cells they attach to the F pilus, and the loss of this pilus is a common reason for a failed or poor transformation/transfection of M13. JM101 is a strain of *E. coli* which possesses the F pilus if the culture is maintained under appropriate conditions. Since the F pilus is not necessary for plasmid DNA transformation, it is advisable to use GM272, a much healthier, F- strain of E. coli for this procedure. To avoid confusion between the similar procedures, bacterial transformation with plasmid DNA is termed a "Transformation", and a bacterial transformation with naked M13 followed by a transfection with intact M13 phage is called a "Transfection."

An additional level of selection can be achieved during transformation and transfections. Bacterial cells containing plasmids with the antibiotic resistance gene are selected in bacterial transformations, and cells in an area of M13 infection are recognized as plaques against a lawn of non-infected cells.

However, the object of most transformations and transfections is to clone foreign DNA of interest into a known plasmid or viral vector and to isolate cells containing those recombinant molecules from each other and from those containing the non-recombinant vector. The E. coli lacZ operon has been incorporated into several cloning vectors, including plasmid pUC and bacteriophage M13. The polylinker regions of these vectors was engineered inside of the lacZ gene coding region, but in a way not to interrupt the reading frame or the functionality of the resultant lacZ gene protein product. This protein product is a galactosidase. In recombinant vectors which have an insert DNA molecule cloned into one of the restriction enzyme sites in the polylinker, this insert DNA results in an altered lacZ gene and a non-functional galactosidase. The presence or absence of this protein can easily be determined through the use of a simple chromogenic assay using IPTG and X-Gal. IPTG is the lacZ gene inducer and is necessary for the production of the galactosidase. The usual substrate for the lacZ gene protein product is galactose, which is metabolized into lactose and glucose. X-Gal is a colourless, modified galactose sugar. When this molecule is metabolized by the galactosidase, the resultant products are a bright blue color.

When IPTG and X-Gal are included in a plasmid DNA transformation, blue colonies represent bacteria harboring non-recombinant pUC vector DNA since the lacZ gene region is intact. IPTG induces production of the functional galactosidase which cleaves X-Gal and results in a blue colored metobolite. It follows that colourless colonies contain recombinant pUC DNA since a nonfunctional galactosidase is induced by IPTG which is unable to cleave the X-Gal. Similarly, for bacteriophage transfections, colourless plaques indicate regions of infection with recombinant M13 viruses, and blue plaques represent infection with non-recombinant M13.

Host Mutation Descriptions

ara Inability to utilize arabinose.

deoR Regulatory gene that allows for constitutive synthesis for genes involved in deoxyribose synthesis. Allows for the uptake of large plasmids.

endA DNA specific endonuclease I. Mutation shown to improve yield and quality of DNA from plasmid minipreps.

F' F' episome, male E. coli host. Necessary for M13 infection.

galK Inability to utilize galactose.

galT Inability to utilize galactose.

gyrA Mutation in DNA gyrase. Confers resistance to nalidixic acid.

hfl High frequency of lysogeny. Mutation increases lambda lysogeny by inactivating specific protease.

lacI Repressor protein of lac operon. LacI[q]is a mutant lacI that overproduces the repressor protein.

lacY Lactose utilization; galactosidase permease (M protein).

lacZ b-D-galactosidase; lactose utilization. Cells with lacZ mutations produce white colonies in the presence of X-gal; wild type produce blue colonies.

lacZdM15 A specific N-terminal deletion which permits the a-complementation segment present on a phagemid or plasmid vector to make functional lacZ protein.

Dlon Deletion of the lon protease. Reduces degradation of b-galactosidase fusion proteins to enhance antibody screening of l libraries.

malA Inability to utilize maltose.

proAB Mutants require proline for growth in minimal media.

recA Gene central to general recombination and DNA repair. Mutation eliminates general recombination and renders bacteria sensitive to UV light.

rec BCD Exonuclease V. Mutation in recB or recC reduces general recombination to a hundredth of its normal level and affects DNA repair.

relA Relaxed phenotype; permits RNA synthesis in the absence of protein synthesis.

rspL 30S ribosomal sub-unit protein S12. Mutation makes cells resistant to streptomycin. Also written strA.

recJ Exonuclease involved in alternate recombination pathways of *E. coli*.

strA See rspL.

sbcBC Exonuclease I. Permits general recombination in recBC mutants.

supE Supressor of amber (UAG) mutations. Some phage require a mutation in this gene in order to grow.

supF Supressor of amber (UAG) mutations. Some phage require a mutation in this gene in order to grow.

thi-1 Mutants require vitamin B1(thiamine) for growth on minimal media.

traD36 mutation inactivates conjugal transfer of F' episome.

umuC Component of SOS repair pathway.

uvrC Component of UV excision pathway.

xylA Inability to utilize xylose.

Commonly Used Bacterial Strains

C600 - F-, e14, mcrA, thr-1 supE44, thi-1, leuB6, lacY1, tonA21, [[lambda]] [-]

-for plating lambda (gt10) libraries, grows well in L broth, 2x TY, plate on NZYDT+Mg.

-Huynh, Young, and Davis (1985) DNA Cloning, Vol. 1, 56-110.

DH1 - F[-], recA1, endA1, gyrA96, thi-1, hsdR17 (rk[-], mk[+], supE44, relA1, [[lambda]][-]

]-for plasmid transformation, grows well on L broth and plates.

-Hanahan (1983) J. Mol. Biol. 166, 557-580.

XL1Blue-MRF'-D(mcrA)182, D(mcrCB-hsdSMR-mrr)172, endA1, supE44, thi-1, recA, gyrA96, relA1, lac, l-, [F'proAB, lac I[q]ZDM15, Tn10 (tet[r])] -For plating or glycerol stocks, grow in LB with 20 ug/ml of tetracycline. For transfection, grow in tryptone broth containing 10 mM MgSO4 and 0.2% maltose. (No antibiotic—Mg2+ interferes with tetracycline action.) For picking plaques, grow glycerol stock in LB to an O.D. of 0.5 at 600 nm (2.5 hours). When at 0.5, add $MgSO_4$ to a final concentration of 10 mM.

SURE Cells - Stratagene - e14(mcrA), D(mcrCB- hsdSMR-mrr)171, sbcC, recB, recJ, umuC::Tn5 (kan[r]), uvrC, supE44, lac, gyrA96, relA1, thi-1, end A1[F'proAB, lacI[q]DM15, Tn10(tet[r])]. An uncharacterized mutation enhances the a[-] complementation to give a more intense blue colour on plates containing X-gal and IPTG.

GM272 - F[-], hsdR544 (rk[-], mk[-]), supE44, supF58, lacY1 or [[Delta]]lacIZY6, galK2, galT22, metB1m, trpR55, [[lambda]][-]

-for plasmid transformation, grows well in 2x TY, TYE, L broth and plates.

-Hanahan (1983) J. Mol. Biol. 166, 557-580.

HB101 - F[-], hsdS20 (rb[-], mb[-]), supE44, ara14, galK2, lacY1, proA2, rpsL20 (str[R]), xyl-5, mtl-1, [[lambda]][-], recA13, mcrA(+), mcrB(-)

-for plasmid transformation, grows well in 2x TY, TYE, L broth and plates.

-Raleigh and Wilson (1986) Proc. Natl. Acad. Sci. USA 83, 9070-9074.

JM101 - supE, thi, [[Delta]](lac-proAB), [F', traD36, proAB, lacIqZ[[Delta]]M15], restriction: (rk[+], mk[+]), mcrA+

-for M13 transformation, grow on minimal medium to maintain F episome, grows well in 2x TY, plate on TY or lambda agar.

-Yanisch-Perron et al. (1985) Gene 33, 103-119.

XL-1 blue recA1, endA1, gyrA96, thi, hsdR17 (rk[+], mk[+]), supE44, relA1, [[lambda]][-], lac, [F', proAB, lacIqZ[[Delta]]M15, Tn10 (tet[R])]

-for M13 and plasmid transformation, grow in 2x TY + 10 ug/ml Tet, plate on TY agar + 10 ug/ml Tet (Tet maintains F episome).

-Bullock, et al. (1987) BioTechniques 5, 376-379.

GM2929 - from B. Bachman, Yale E.coli Genetic Stock Center (CSGC#7080); M.Marinus strain; sex F[-];(ara-14, leuB6, fhuA13, lacY1, tsx-78, supE44, [glnV44], galK2, galT22, l[-], mcrA, dcm-6, hisG4,[Oc], rfbD1, rpsL136, dam-13::Tn9, xyl-5, mtl-1, recF143, thi-1, mcrB, hsdR2.)

MC1000 - (araD139, D[ara-leu]7679, galU, galK, D[lac]174, rpsL, thi-1). obtained from the McCarthy lab at the University of Oklahoma.

ED8767 (F-,e14-[mcrA],supE44,supF58,hsdS3[rB[-]mB[-]], recA56, galK2, galT22,metB1, lac-3 or lac3Y1 , obtained from Nora Heisterkamp and used as the host for abl and bcr cosmids.

Notes on Restriction/modification Bacterial Strains

1. EcoK (alternate=EcoB)-hsdRMS genes=attack DNA not protected by adenine methylation. (ED8767 is EcoK methylation minus).
2. mcA (modified cytosine restriction), mcrBC, and mrr=methylation requiring systems that attack DNA only when it IS methylated (Ed8767 is mrr+, so methylated adenines will be restricted. Clone can carry methylation activity.)
3. In general, it is best to use a strain lacking Mcr and Mrr systems when cloning genomic DNA from an organism with methylcytosine such as mammals, higher plants , and many prokaryotes.
4. The use of D(mrr-hsd-mcrB) hosts=general methylation tolerance and suitability for clones with N6 methyladenine as well as 5mC (as with bacterial DNAs).
5. XL1-Blue MRF'=D(mcrA)182, D(mcrCB-hsdSMR-mrr)172,endA1, supE44, thi-1, recA, gyrA96, relA1, lac, l-, [F' proAB, lacI[q]ZDM15, Tn10(tet[r]

pUCl8 is an plasmid, its basic structure is shown in Figure 41. Plasmid pUCl8 is a small circular DNA molecule that contains only 2686 nucleotide pairs (molecular weight = 2xl06). Small plasmids generally replicate more efficiently in bacteria and produce larger numbers of plasmids per cell. As many as 500 copies of this plasmid may be present in a single *E . coli* cell. Plasmid pUC18 contains an ampicillin-resistance gene that enables *E . coli* to grow in the presence of the antibiotic. Bacteria lacking this plasmid, or bacteria that lose the plasmid, generally will not grow in the presence of this antibiotic. The ampicillin resistance gene of pUC18 codes for the enzyme beta-lactamase (penicillinase) which inactivates ampicillin and other penicillins.

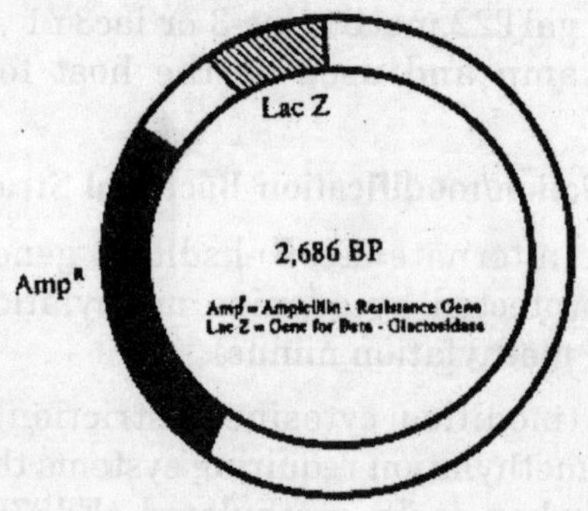

Fig. 41: Partial map of plasmid pUC18

In the laboratory, plasmids can be introduced into living bacterial cells by a process known as transformation. When bacteria are placed in a solution of calcium chloride, they acquire the ability to take up plasmid DNA molecules. As illustrated in Figure 42, this procedure provides a means for preparing large amounts of specific plasmid DNA since one transformed cell gives rise to a clone of cells that contains

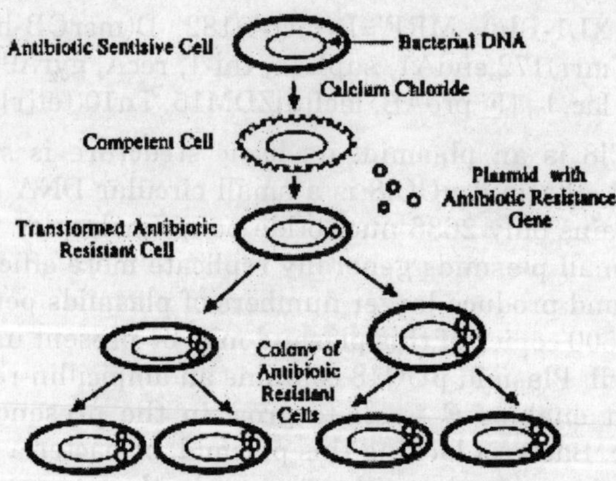

Fig. 42: Transformation of bacteria with plasmid DNA

exact replicas of the parent plasmid DNA molecule. Following growth of the bacteria in the presence of the antibiotic, the plasmid DNA can readily be isolated from the bacterial culture.

Plasmids are very useful tools for the molecular biologist because they serve as gene-carrier molecules. A basic procedure of recombinant DNA technology consists of joining a gene of interest to plasmid DNA to form a hybrid, or recombinant molecule that is able to replicate in bacteria (Figure 43). In order to prepare a recombinant molecule, the plasmid and gene of interest are cut at precise positions by specific deoxyribonucleases (restriction endonucleases) and then the molecules are spliced together. After the hybrid plasmid molecule has been prepared, it is introduced into *E. coli* cells by transformation. The hybrid plasmid replicates in the dividing bacterial cells to produce an enormous number of copies of the original gene. At the end of the growth period,

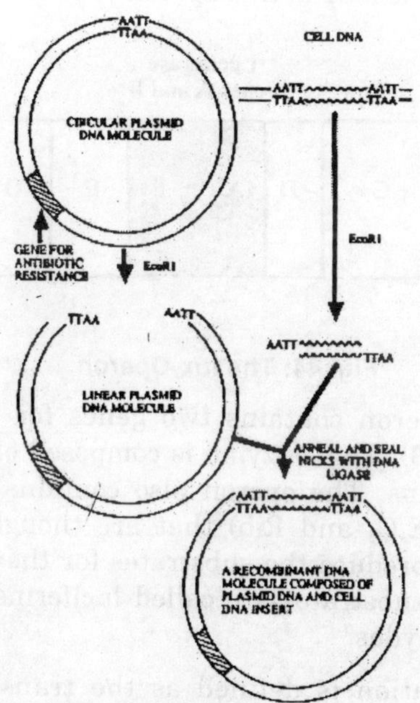

Fig. 43: Formation and amplification of a recombinant DNA molecule

the hybrid molecules are purified from the bacteria and the original gene of interest is recovered. This method has enabled scientists to obtain large quantities of more than 1000 specific genes including the genes for human interferon, insulin and growth hormone.

The Lux Operon

The emission of light by living organisms is a fascinating process. Luminescent bacteria are the most abundant and widespread of the luminescent organisms found in nature. The genes responsible for light emission in a few of these organisms have been well characterized. The genetic system required for luminescence in the bacterium Photobacterium (Vibiro) fischeri is the lux operon. This operon contains two genes that code for luciferase (the enzyme that catalyzes the light-emitting reaction) and several genes that code for enzymes which produce the luciferins (which are the substrates for the light-emitting reaction). A genetic map of this operon is shown in Figure 44 .

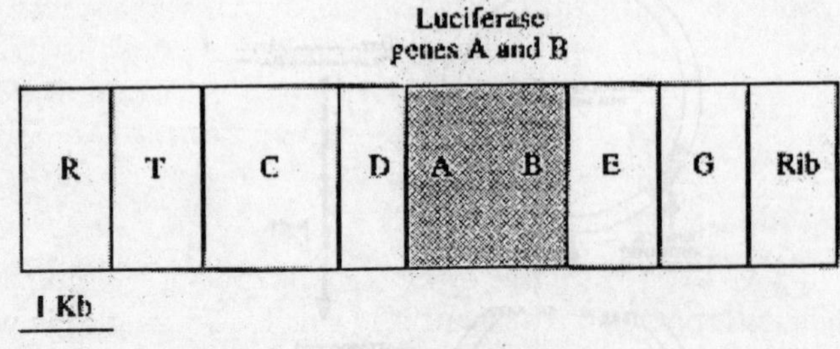

Fig. 44: The lux Operon

The lux operon contains two genes for the luciferase enzyme (A and B). This enzyme is composed of two different polypeptide chains. The operon also contains several other genes (R,T,C,D,E,G, and Rib) that are thought to code for enzymes which produce the substrates for the light-emitting reaction. These substrates are called luciferins and are long chain fatty aldehydes.

Transformation is defined as the transfer of genetic information from a donor to a recipient using naked DNA. The recipient takes DNA up from the media, and there is no requirement for cell to cell contact.

While naturally transformable bacterial strains exist, these are sufficiently rare that induced transformation is more important to the geneticist. Induced transformation is the act of tricking a normally non-transformable bacteria into taking up DNA from the media. It involves two hurdles: first, the purification of DNA which can be either plasmid or chromosome, linear or circular; secondly, the generation of "competent" cells, which are cells which can accept DNA. In the case of *E. coli*, the following considerations apply: (i) you need multivalent cations and some time of incubation at low temperature; (ii) only a fraction of the cells are competent (<10%); (iii) there are two phases, DNA uptake and establishment, and the latter seems limiting; (iv) there appear to be approximately 100 "channels" per cell for DNA uptake.

Alternatively, Gram-positive bacteria can be incubated with degradative enzymes to remove the peptidoglycan layer and thus form protoplasts. When these latter cell forms are incubated with DNA and polyethylene glycol, one obtains cell fusion and concomitant DNA uptake. In both of these examples, if the DNA is linear, it tends to be very sensitive to nucleases so that transformation is most efficient when it involves the use of covalently closed circular DNA. Alternatively, nuclease-deficient cells (RecBC- strains) can be used to improve transformation.

Finally, the technique of electroporation is coming into more use with bacteria. It essentially involves fusing cells with pulses of electric current and can be used for both cell fusion and incorporation of DNA. When optimized for a particular organism, electroporation seems to be good for at least an order of magnitude increase in the frequency of transformants when compared to optimized transformation methods.

There are also cases of natural transformation where bacterial species naturally take up DNA from the environment: *Bacillus, Streptococcus pneumoniae, Haemophilus,* and *Neisseria.* With these bacteria, the competence to accept DNA is natural, but often is dependent on the cells being in a particular growth stage. The DNA binding in these bacteria is by particular receptors and, in the case of some of the gram negative cells, particular DNA sequences are required. In

Gram-positive bacteria, the process is exceedingly complex and the DNA is cut into fragments and converted to single strands which are protected by DNA binding proteins. Integration into the chromosome is apparently by single-strand invasion of the DNA duplex followed by ligation. In this latter case, plasmids are poorly transformed since they lack homology with the chromosome. This can be solved either using a recipient with a homologous plasmid or using plasmid multimers so that the overlapping single strand fragments can regenerate a plasmid monomer.

In general, transformation allows the transfer of reasonably small, apparently random, pieces of chromosomal or plasmid DNA from a donor to a recipient. As with any gene transfer system, the transferred DNA is selectable only if stably inherited in the recipient cell. This is done either by introducing genes that themselves can replicate (which requires both the ability to circularize and to be associated with a replicon) or can integrate into or recombine with a replicon in the recipient cell. Compared with conjugation, transformation is rather slow and work-intensive, not only because of the requirement for isolation of DNA, but also because competent cells need to be produced. However, it is often the only method available, especially if the DNA in question is a result of in vitroconstructions. Transformation also tends to become extremely inefficient as the size of the DNA increases. One of the solutions has been to package large pieces of DNA into phage heads in vitroand then have the constructed phage inject the DNA into the desired cells.

Transduction: Transduction is defined as the transfer of genetic information between cells through the mediation of a virus (phage) particle called bacteriophage. It therefore does not require cell to cell contact and is DNase resistant. There are a wide variety of bacteriophages, those can mediate genetic transfer, these bacteriophages contain double stranded DNA. These phages can be broken down into two classes, both of which are useful for genetic analysis, and these are termed lytic and temperate (or lysogenic).

When a lytic phage injects its DNA into a host cell, it begins an ordered sequence of gene expression that involves the generation of many copies of the phage DNA, typically in the form of long concatamers (a linear sequence of covalently attached phage genomes). The phage also direct the generation of proteinaceous "heads" and then pack their DNA into these heads by some mechanism. When "mature" phage are assembled, they cause the cell to lyse and release approximately 102 to 103 progeny. Another class of phage, termed temperate, have a choice of "lifestyles" upon injecting their DNA into the host. Under some conditions, they follow a route similar to that chosen by lytic phage sketched above. However, they also possess the ability to repress their lytic functions and take up semi-stable residency in the host. This residency typically takes the form of an integration of the phage DNA into the host chromosome, usually by a site-specific recombination event with the resulting integrated phage termed a prophage. This integration event is reversible and is driven largely by phage factors in both directions. If a prophage decides to lyse the cell, it excises itself from the chromosome, begins expression of DNA synthesis and other lytic functions, packages its DNA appropriately, and lyses the cell.

There is another class of *E. coli* phage that possess a number of qualities useful for molecular genetics, but cannot mediate transduction. These are the single-stranded DNA phages (M13 is the best known example) that are filamentous and produce progeny without lysing the host cell. The first feature has the effect that there are not strong size constraints on inserts into these phage, since they do not utilize a "headfull packaging" scheme. The latter property allows very high titres of phage to accumulate in the medium without significant amounts of cell debris. Their single-strandedness is also useful in sequencing strategies and as a substrate for site-directed mutagenesis in vitro. The phage can also be used as "helpers" to cause single-stranded replication and extrusion of plasmids that have been engineered to contain the phage origin of replication.

The production and quantitation of phage stocks requires the exposure of phage to actively growing bacteria. The optimum production of a high-titre lysate (greater than 109 phage/ml) requires inoculation of a bacterial culture with phage under conditions where the phage just lyses the culture as the cell concentration reaches saturation. Optimal levels of bacterial and phage inocula are a function of the growth rate of the bacteria, the latent period (the time from phage infection until cell lysis), and the burst size of the phage. If purified phage (free of cells and debris) are desired, this is usually accomplished by a centrifugation step where the phage position themselves in a salt gradient according to their buoyant density. Phage concentrations can be titred (counted) by plating by various dilutions of the phage stock onto a soft agar lawn containing susceptible cells. The dilution that yields a countable number of plaques, each arising from a single phage, is used to calculate the concentration of the original phage stock.

This DNA transfer mechanism has been enormously useful for *E. coli*, *S. typhimurium* and *B. subtilis*, but such phage have not been identified and developed for most bacteria.

The most commonly used generalized transducing phages, P22 (infecting *S. typhimurium*) and P1 (infecting *E. coli*), package by "headfull packaging", so that they fill the head with a little more than one phage length of DNA. They typically recognize their own DNA as appropriate for packaging because of special sequences termed pac sites. Occasionally, the phage make an error and begin to package a host "concatamer", presumably because a site on the host DNA is reminiscent of the phage pac site. It is also conceivable that insertion sequences in the host might cause a transposition of the host DNA into the phage concatamer, thus allowing packaging of host DNA even when the packaging scheme started correctly on phage concatamers.

For both P22 and P1, about one phage virion in 1,000 actually contains host DNA. It is possible to obtain mutants which make errors more frequently. An example of this is the mutant termed P22HT which has completely lost the ability to recognize its pac sites so that it packages DNA randomly. Since an infected cell contains about half host and half phage DNA, the P22HT variant causes the generation of a phage lysate in which half of the phage heads contain host DNA. In the normal case, to the extent that mispackaging is due to pseudo-pac sites in the host DNA, one would expect that not all regions of the host DNA would be transduced at equal frequency. Such seems to be the case with wild-type P22 and this differential transduction is eliminated in mutants like P22HT.

As always, for the transferred DNA to become stably inherited, it must become associated with a replicon in the recipient cell. Since it is a linear molecule, this requires two homologous recombination events with the net result that a portion of the chromosome in the recipient is replaced by a portion of the incoming DNA. Thus, in contrast with specialized transduction, generalized transduction requires that the recipient possess both a functional rec system and DNA homologous to the transferred fragment. It has been observed in generalized transduction that there is an approximately 10% chance of the incoming chromosomal DNA being recombined into the recipient's chromosome. A recipient cell whose genotype and phenotype have been changed by a transduction event is called a transductant.

The two phage referred to above carry 1-2% of the host DNA and normally package chromosomal DNA at a frequency of 10-3. Thus, about one in 105 phage particles carries a given region of the host DNA. Since a given gene injected into a recipient cell by a transducing particle has a 10% chance of being recombined into the chromosome then one might expect to get approximately one transductant for a given region per 106 phage particles. In the case of the P22HT derivative, since approximately half of the phage particles carry chromosomal DNA, one can achieve the extraordinary frequency of one transductant per 1000 phage particles. There is also a remarkable phage in *B. subtilis* that carries 10% of the host chromosome, though apparently its frequency of mispackaging is sufficiently low that the number of transductants per phage particle is around one in 107. In general, the virtue of generalized transduction is not its efficiency, but its ease of use. For strain construction, it is also an advantage that only a portion of the chromosome is transferred, in contrast to Hfr's.

Specialized phage are derivatives of temperate phage where some portion of the phage genome has been replaced by a portion of other genetic material, typically the bacterial chromosome. In this way, all of the phage particles carry the same portion of the bacterial chromosome. If one considers the structure of a prophage as indicated below, and asks what would happen if the prophage came out imprecisely, the result would be a specialized phage. This phage would have the property that it would be mutant in some way (although they may still possess all functions necessary for growth; if they fail to grow without helper phage, they are termed "defective") as some of the phage functions had been effectively deleted; and all progeny would carry the same host sequence. Typically one solves the difficulty of the defectiveness of the phage by supplying a normal wild-type version of the phage, termed a "helper" which grows along with a specialized phage and supplies whatever functions are necessary for generating phage particles. Since it would not be very interesting to pick up only those genes which happen to be located on either side of the phage att site, other genes can be put on specialized phage by several methods including (a) forcing the prophage to go in at alternate sites, (b) moving the gene of interest near a phage attachment site, (c) cloning the gene of interest onto the specialized phage. While the first two methods require little understanding of the physical structure of the phage, the last method requires a good deal of that information. One needs to have at least one restriction sequence that occurs only once in the phage (and in a non-essential region) to serve as a site for cloning. Fortunately, through the diligent efforts of molecular biologists, there are a plethora of phage that have all the desired properties, at least for *E. coli*. Since phage can only carry a "headfull" of DNA, a region of non-essential DNA needs to be removed from the phage genome to compensate for that inserted DNA. Alternatively, a helper phage can be used as noted above. Finally, it can often be useful to have a drug-resistance gene cloned into the phage to serve as an easy assay for the genetic presence of the phage. Specialized phage have two possible ways of introducing genes into a recipient. The first and most efficient involves the normal phage integration system. By this means, the entire specialized phage

DNA is integrated at the phage attachment site, possibly in conjunction with the wild-type helper phage. This generates a merodiploid (a strain partially diploid for a portion of its genome) for the region carried by the phage. The other possibility involves recombination between the chromosomal DNA carried by the phage and the chromosome itself.

As always, for a transferred gene to be "stably" inherited in the recipient, it needs to be replicated in that recipient. In the case of transduction, this means that the incoming DNA must (generally) integrate into a replicon in the recipient (either the chromosome or a plasmid). Transferred DNA in generalized transduction can only do this by homologous recombination while DNA associated with specialized phage also has the phage int system as a means of association with the replicon. Other "exceptions" to these rules are (i) transposons that can transpose into the chromosome of a Rec+ or Rec- cell from DNA of a specialized or generalized transducing phage; (ii) a specialized phage that has the ability to circularize and replicate as a plasmid in the recipient cell, thus not requiring any recombination into the chromosome; (iii) a plasmid that is moved by generalized transduction or by transformation and that has the ability to circularize and subsequently replicate in the recipient.

Transposition: There are two types of transposition:

Conservative Transposition: Conservative transposition is so named because the copy number of the transposon is conserved during the operation. It involves the transposon physically leaving one replicon and moving to another. The original replicon is apparently destroyed so it is critical that the element transpose to another replicon in the cell. Most elements try to transpose soon after DNA replication because there is then the highest probability that the cell actually contains another replicon. Notice that while conservative transposition does not increase the element's copy number in the cell (at least not initially), it does increase the ratio of transposons to replicons in the cell. This might actually be the reason that Tn's do not repair the site they have left. The model was developed for Tn5/IS50 and seems to describe the

Tn10/IS10. The initial events are (a) double-strand breaks on both sides of the transposon coupled with staggered single-strand breaks in the target replicon. (b) The ends of the transposon are then ligated to the single-strand segments in the target DNA. Finally, the gaps in the target molecule are filled in.

Bi-directional Replicative Transposition: This model differs from the previous one in several important ways, the transposon increases its copy-number in the cell directly and the donor molecule is not destroyed. It apparently explains the mechanism employed by Tn3 and Mu. It involves.

1. Single-strand breaks in the target DNA and on either side of the transposon
2. Ligation of ends as shown to produce a complicated hybrid molecule
3. bi-directional replication through the element with subsequent ligation
4. This results in production of a single replicon where the donor replicon is flanked by two copies of the transposon and fused to the target replicon.
5. A replication event "resolves" this to the finished product

In the case of each of these models it should be realized that they are in fact only models and definitive proof of these mechanisms is still lacking, though the recent development of in *vitro* transposition systems is extremely promising.

Transposon

A transposon is a mobile genetic element containing additional genes unrelated to transposition functions. In general, there are known to be two general classes: (1) Class-1 and (2) Class-2

Class 1 or "compound Tns" encode drug resistance genes flanked by copies of an IS in a direct or indirect repeat. A direct repeat exists when the two sequences at either end are oriented in the same direction while an indirect (or inverted) repeat exists when they are in opposite directions. In this class of transposons, the IS sequence supplies the transposition function. These are essentially IS's, where in two copies of the IS have flanked another gene. Examples are Tn5, 9, 10, 903, and 1681. In all of these, the ends of the transposon are capable of transposing separately, since they are in fact still IS's.

The second class of transposons are known as "complex" or Class 2. With these, the element is flanked by short (30-40 bp) indirect repeats with the genes for drug resistance and transposition encoded in the middle . Not surprisingly, these must always transpose as a unit. Examples are Tnl, 3, 4, 7, 501, and 551. The following sections will consider five general properties of transposons, namely transposition, its regulation, polarity, deletion generation, and precise excision. These properties will be treated first for Tn10, which is an example of the Class l transposons, and then with Tn3, an example of Class 2 transposons.

Class 1 Transposons (Specifically Tn10 Which Is Flanked By Two Copies Of Is10)

Transposition: Transposition of IS10 or Tn10 occurs at around 10-7 per element per generation. Transposition occurs to a large number of sites in E. coli but excellent target sites are found at about 1 per 1000 base pairs. Other, secondary, sites are recognized at poor efficiency. The preferred sites for IS10 and Tn10 tend to have the sequence GCTNAGC (admittedly not very AT-rich) which is then duplicated within the 9 base pair duplication upon insertion.

Regulation of Transposition: Kleckner has found that transposition is regulated by a large set of factors and circumstances. These all serve to reduce the level of transposition: (a) There are two RNA's produced from one end of the IS10, one which encodes the transposase and the other which is non-coding. These two promoters are separated by 33 base pairs and when the RNA's are both being synthesized and can hybridize to each other, the presence of the second RNA inhibits the translation of the transposase encoded by the first. This second RNA is therefore a trans-acting inhibitor of transposase expression. The Pout promoter is also much stronger than Pin. (b) The coding mRNA appears to be relatively unstable and poorly translated, even without this complementary RNA. (c) Transposase is essentially cis-acting, so it does not float around to cause transposition of other copies of the transposon. (d) Transposase has a dramatic preference for hemi-methylated (due to the dam system) ends of the transposon, resulting in a 104-fold increase in transposition following replication through the element. This restricts transposition both to a very short time frame and to a period where there are at least two replicons in the cell. (e) The transposase functions as a multimer, so single proteins are ineffective at causing transposition.

Polarity: Tn10 is polar on downstream genes, but can turn on downstream genes if and only if the strain is deficient in the function Rho or if all Rho-dependent sites are removed between the site of the transposon and the gene in question. The promoter that allows the activation of these downstream genes is Pout.

Chromosomal Deletions: With Tn10, it is possible to demand a loss of the material between the flanking IS10 elements because a selection exists for tetracycline sensitivity (Tn10 confers tetracycline resistance). A variety of events are detected but typically one finds deletions which start at the IS10 and delete all the intervening information between that and the other IS10. Such events are often accompanied by inversions of contiguous DNA sequences beginning at one of the IS10 ends.

Precise Deletion: Precise deletion of IS10 or Tn10 occurs at approximately 10-8 per element per generation. This seems to be independent of the recA system and may reflect a "copy-choice" mechanism in replication. Another event, termed "nearly precise", causes the deletion of all but 50 base pairs near the ends of the IS elements. These events occur at around 10-6. Surprisingly, at the frequency of 10-6, these imprecise deletions can give rise to a precise deletion event (a return to the wild-type genotype).

Class 2 or Complex Transposons (Using Tn3 as the Example)

Transposition: Transposons of this class tend to show regional specificity and Tn3 in particular seems to especially prefer AT-rich regions that have some homology to its ends. Unlike Class l, the ends cannot transpose separately. One member of this class, Tn7 has such extreme specificity that it has a single site of preference in all of the E. coli chromosome. It has now been shown that the region conferring this specificity is not the site of insertion, but is apparently nearby. A similar separation of recognition and "action" sites has been seen with certain restriction enzymes. The frequency of transposition tends to be variable ranging from 10-4 to 10-6 and, in the case of Tn3, involves a five-base-pair duplication upon insertion.

Regulation of transposition: The structure of Tn3 is indicated in the accompanying figure. What should be noted is that the products of two genes, tnpA and tnpR, are involved in some way in the transposition event. The product of tnpR, the resolvase, is also a regulatory protein, acting as a repressor of both its own synthesis and tnpA, which encodes the transposase. The mode of regulation of other transposons of this class is unknown. Tn3 has also been shown to possess an "immunity" function that reduces the frequency of other Tn3 insertions into the same replicon. The mechanism of this immunity is unknown.

Polarity: The situation with this class of transposons in terms of their polarity is somewhat uncertain. It is quite clear, however, that in one orientation at least, these transposons tend to be polar. It is possible that the lack of polarity in other cases is due to the initiation of new transcripts from transposon promoters and not due to reading through across the transposon.

Deletion Generation: Both deletion and inversion events next to the transposon are frequent. The end points of both the deletions and the inversions seem to be non-random and in the case of inversions, there is typically a second copy of the transposon at either end of the inverted region.

Element Deletion: Tn3 does not appear to be deleted precisely at a detectable frequency. Even in those few cases where revertants to a wild-type phenotype occur, subsequent analysis has shown that the wild-type genotype has not been restored. Such events that might restore a wild-type phenotype without the wild-type genotype will be discussed in the section on suppressors near the end of the text. For another transposon of this class, Tn101, reversion to a wild-type genotype has been shown to occur at 10-11, which is essentially undetectable.

Bacteriophage Mu: Bacteriophage Mu belongs in the general category of a class-2 transposon. Its physical size is 38 kb and it generates 5 base pair duplications upon insertions. It produces an 11 base pair inverted repeat at either end. Its site preference is remarkably random and the argument has been made that its specificity can be for no more than one or two base pairs. However, in at least one particular region, it has been found that a disproportionate number of insertions fall within one very small region of the gene suggesting that there can be some site preference. Mu is rather strongly polar in both orientations, but it is clear that there is an exceedingly low level of transcription out of one end of the prophage. The transposition of Mu is known to generate deletions as roughly 10% of the Mu prophages have adjacent deletions. These deletions tend to start at one end or the other of the prophage and extend into the adjoining DNA though there also seem to be cases where the deletions are unlinked to the prophage. Finally, precise deletion of Mu is rather rare, occurring at approximately 10-9, and seems to be dependent upon at least some Mu factors. The advantages of the use of Mu are: it is not normally found in the bacterial genome and therefore there are few problems with homology to existing sequences in the chromosome; in contrast to most other transposons, Mu does not need a separate vector system, since it is itself a vector, being a bacteriophage; Mu prophage (at least the cts versions, where c encodes the repressor) are inducible. The disadvantage of Mu is that it is a bacteriophage and therefore can kill the host cell. A wide variety of useful mutants of Mu have been generated.

Chapter—6 | Bacterial Diseases of Plants

Introduction

Plants are also suffered from bacterial, viral or fungal attacks like human being. The organisms themselves (pathogen) are different, but at the microbial level the infection is much the same since one cell is as good a host as another. It is not a matter that which part of the plant is affected because the effect is usually to weaken or kill the whole plant. By infecting the leaves the plant's ability to produce its food is reduced. Some pathogenic bacteria block the vessels in the stems which supply the leaves and by attacking the roots, the uptake of water and nutrients is reduced or stopped completely. Sometimes the bacterial infection is symbiotic where both organisms derive a benefit. Nitrogen fixing bacteria is a suitable example for this. It resides in nodules on the roots of leguminous plants, here the plant provides food and protection, the bacteria takes nitrogen from the air and converts it to a form usable by the host. While most bacteria in the environment are beneficial, but several are able to cause diseases like leaf spots, stem rots, root rots, galls wilts, blights and cankers.

Burrill (1878) was the first to describe the association of bacteria with plant diseases and Smith by 1900 established the role of bacteria as plant pathogens. Plant pathogenic bacteria generally survive in infected plants, in debris from infected plants, and in a few cases, in infested soil. Bacteria are incapable of mechanically penetrating the cutinized plant tissues, cuticle, periderm etc. Most require a wounds (caused by insects, nematodes or other micro organisms) or natural opening in the plant like non-cutinized areas (root hairs, stingmas etc). and stomata, lenticel to gain entry and require warm, moist conditions in order to cause disease. On entry into the host, they colonize intercellularly, intracellularly or intravascularly. Water is a major factor in the spread and development of bacterial diseases. Bacteria grow between plant cells on the nutrients that leak into that space or within the vascular tissues of the plant. Depending on the species of bacteria involved and the tissue infected, they release enzymes that degrade cell walls, toxins that damage cell membranes, growth regulators that disrupt normal plant growth, and complex sugars that plug water conducting vessels. In most bacterial diseases, photosynthesis and respiration are severely altered to the detriment of the plant. The bacterial diseases may be soil borne, seed borne or vector borne. Their spread by air is not significant because the cells lose their viability when carried by winds. Bacteria reproduce very rapidly. They are splashed easily from the soil to the leaves and from leaf to leaf by overhead irrigation. Approximately 170 species of bacteria can cause disease on foliage plants. Bacteria are normally present on plant surfaces and will only cause problems when conditions are favourable for their growth and multiplication. These conditions include high humidity, crowding and poor air circulation around plants. Misting plants will provide a film of water on the leaves where bacteria can multiply. Too much, too little, or irregular watering can put plants under stress and may predispose them to bacterial infection. Other conditions that produce stress include low light intensity, fluctuating temperatures, poor drainage, too small or too large a pot, and deficient or excess nutrients. They are also easily moved from soil or debris when a worker handles such material and then handles the live plant. The most important means of avoiding crop specially ornamental crop losses caused by bacteria is to purchase plants that have been shown to be free of such pathogens by the process of culture indexing. In this procedure pieces of plant tissue are incubated in a nutrient broth which will encourage the growth of plant pathogenic bacteria. If the test is repeated two to three times and no pathogenic bacteria are detected, the plant is said to have been indexed and free of bacterial pathogens.

In contrast to viruses and some obligate fungal pathogens, all bacteria, pathogenic or nonpathogenic, are saprophytes and can be cultured on artificial media. No obligatory pathogenic bacteria are known so far. None of the plant pathogenic bacteria have been found to infect man and animals and *vice versa*. All the plant pathogenic bacteria are rodshaped. A majority of the phytopathogenic bacteria are flagellated. Among the major plant pathogenic bacterial genera *Pseudomonas, Xanthomonas Erwinia* and *Agrobacterium* are gram-negative while the remaining genera *Clavibacter, Curtobacterium, Rhodococcus* and *Streptomyces* are gram-positive.

Common Symptoms of Bacterial Plant Diseases

The bacterial plant diseases produce different types of symptoms depending upon the species, host parasite relationship and environments. The following are the major types of plant disease symptoms commonly observed to be due to bacterial pathogens.

1. Local Lesions

Bacterial lesions or localized spots occur on the leaf blade and sometimes in the petiole, stem, etc. Such spots may also occur on fruits, resulting in the discolouration of fruits. The leaf spots start as minute watersoaked specks, spreading rapidly to form circular, irregular or angular spots, bound by veins and veinlets. The pathogen which enters the hosts tissues through stomatal openings establishes in the substomatal space and remains so in the parenchymatous tissues causing necrotic lesions. This necrosis may spread rapidly in certain cases, such as in fire blight of apples and pears.

2. Soft Rots

The soft rot symptoms are found mainly on fleshy parts. The major effect is the softening of tissues due to disintegration of cells and dissolution of middle lamella as result of the action of enzymes and very often a dirty liquid oozes out of the affected parts. The enzymes diffuse in advance to dissolve the middle lamella of the cells. Plasmolysis and death of cells follow and the bacteria grow upon the dead plant tissues rather than upon the living cells.

3. Vascular Disease

In some of the bacterial leaf spots the organism moves into the vascular tissues and becomes systemic. In others, the invasion is concentrated in the vascular tissues, causing typical wilt of the affected plants. Due to vascular infection there may be hypoplasia, leading to the stunting of the entire plant. Common examples are cucurbit bacterial wilt, brown rot of potato, ring rot of potato bacterial wilt of tomato, etc. Certain polysaccharides produced by phytopathogenic bacteria *in vitro* possess the wilt inducing properties. In other studies it has been found that certain pectinolytic and cellulolytic enzymes possess the wilt inducing properties. Pathogenic infection causes the movement of water in the vessels to decline due to an increase in the viscosity of vessel fluid. This is because of the formation of a slimy or gummy substance surrounding the bacterial mass. Tylose formation and vessel collapse may also contribute to wilting.

4. Tumours and Galls

In many bacterial diseases, the effect of invasion by the pathogen is the hyperplasia and hypertrophy of invaded tissues. As a result, tumours develop on the affected organs. The formation of root nodules in leguminous roots in response to bacterial action has also a similar effect. Crown gall (*Agrobacterium tumefaciens*) is the best known example of this. The pathogen seems to trigger a chain of reactions to produce some chemicals in the vicinity which cause hyperplasia and hypertrophy of the host cells.

5. Scabs and Cankers

Scabs and cankers are corky outgrowths which are formed on leaves, twings and all other plant parts above the ground. These outgrowths are the result of the reactions of the host tissues to the pathogen. Such reactions are mostly localized and often confined to the parenchymatous tissues of the host plant. The incubation period normally required for such infections is about ten days after infection. Scab is formed by epidermal infection and is not deep-seated. The powdery scab of potatoes is a familiar example. Cankers, on the other hand, are deep-seated and involve the cambium layer. They entail the destruction of woody tissues and are common infections on stems of citrus, mango and many fruit and forest trees.

Common Control Measures

Since killing the pathogens is difficult or impossible, "prevention is better than cure." By observing good hygiene when propagating and growing plants, it can prevent a lot of diseases from taking following steps:

- Destroy diseased plants, clear up dead leaves and other debris;
- Prune fruit trees and bushes regularly to keep an open structure allowing a good air flow and removing damaged branches;
- Disinfect secateurs and knives used for cutting out diseased branches with methylated spirits or a flame (a cigarette lighter comes in handy for this). Also helps when taking cuttings;
- Only use new or well-washed containers when growing cuttings and sowing seeds;
- Crop rotation in the vegetable plot will prevent a build up of disease;
- Space plants well apart especially crops where similar plants are growing together, to allow good air flow;
- Catching disease early is important so keep an eye out for it at all times;
- Soil used in potting should be treated to kill all pathogens. Soil in which infected plants were grown or rooted should be discarded or thoroughly treated;
- Workers should be trained to not handle soil or debris and then the living plant tissue unless they stop work immediately and wash their hands;
- Do plant handling procedures and debris/soil handling operations completely separately;
- The most important cultural practice used against bacteria is irrigating in a manner that keeps foliage surfaces dry and which avoids splashing. Overhead irrigation should not be used in crops particularly susceptible to bacterial diseases. When overhead watering is employed, watering should be done early in the day so that free moisture evaporates quickly;

- Various types of trickle irrigation and capillary mat watering are techniques that avoid providing the conditions required for bacterial spread and infection;
- Some bacteria have been shown to spread in ebb and flow systems. Steps should be taken to filter crop debris out of the water and chemically treat the water.

In spite of the above precautions, bacterial diseases can control using chemical treatment methods. Though once disease begins on the plants, chemical control is not effective. Although research reports may indicate 80 to 90% control with chemicals under experimental conditions, often less than 50% control is achieved under commercial conditions with chemicals. Plant bacterial diseases can be controlled with the same kinds of antibiotics that are used to control animal diseases, such as streptomycin. Antibiotics are substances which are produced by micro-organisms and which at against microorganism. The chemical nature of antibiotics is complex and is not, as a rule, related to each other. A large number of antibiotics have been tried for plant disease control but only a few have been successfully utilized. The important antibiotics for control of bacterial plant diseases are Streptomycin, Blasticidins, Tetracyclines and Agrimycin-100. Some of the important plant pathogenic genera and their plant diseases have been described in this chapter.

I. Xanthomonas Species and Their Plant Diseases

Xanthomonas have single cells and a single polar flagellum. They are straight rod shaped measuring usually 0.4 × 1.0.4 μ. They are gram-negative bacteria. Their another characters are: non-spore forming; copious extracellular slime produced; strict aerobes; metabolism respiratory, never fermentative; oxidase negative and catalase positive; bacterial colonies yellow coloured due to production of xanthomonadin pigment. *Xanthomonas* have total 100 species, of which all are plant pathogenic. There are some important bacterial diseases caused by *Xanthomonas* are described below:

1. Name of disease: Bacteria blight of cotton or Angular leaf spot or Black arm of cotton

Table—3. Important bacterial pathogens and their plant diseases

Genus Name of bacterial pathogen	G^+/G^-	Plant diseases
I. *Xanthomonas*	G^-(Gram negative)	1. Bacterial blight of cotton
		2. Citrus canker
		3. Bacterial leaf blight of rice
		4. Bacterial leaf streak of rice
		5. Bacterial spot of tomato
		6. Black rot of crucifers
		7. Leaf spot of mango
		8. Bacterial blight of beans
		9. Bacterial blight of ornamental plants
		10. Bacterial leaf spot of foliage plants
		11. Bacterial leaf blight of foliage plants
II. *Pseudomonas*	G^-(Gram negative)	1. Bacterial speck of tomato
		2. Bacterial brown rot of potato
		3. Red stripe of sugarcane
		4. Wild fire of tobacco
		5. Halo blight of beans
		6. Brown spot of beans
		7. Bacterial canker in prunus
		8. Leaf spot of *Dracaena sanderana*
		9. Leaf spot of ornamental plants
		10. Bacterial leaf spot of foliage plants

(Contd...)

Table—3. (Contd...)

Genus Name of bacterial pathogen		*G^+/G^-*		*Plant diseases*
III.	*Erwinia*	G^-(Gram negative)	1.	Bacterial stalk rot of maize
			2.	Black leg and soft rot of potato
			3.	Bacterial wilt of cucurbits
			4.	Fireblight of apple and pears
			5.	Stem rot of ornamental plants
			6.	Bacterial blight of foliage plants
			7.	Rapid decay in Epiprermum
IV.	*Agrobacterium*	G^-(Gram negative)	1.	Crowngall of stone fruits
V.	*Clavibacter*	G^+ (Gram positive)	1.	Bacterial ear rot of wheat
			2.	Ratoon stunting disease of sugarcane
			3.	Bacterial canker in tomato.
VI.	*Curtobacterium*	G^+ (Gram positive)	1.	Bacterial wilt of beans
VII.	*Rhodococcus*	G^+ (Gram positive)	1.	Abnormal branching of ornamental plants
VIII.	*Streptomyces*	G^+ (Gram positive)	1.	Common scab of potato
IX.	*Ralstonia solanacearum*		1.	Vascular wilting of ornamental plants.

(i) **Causal Organism:** Xanthomonas campestris pv malvacearum (Smith) Dye or

Xanthomonas malvacearum (E.F. Smith) Dowson.

(ii) **Symptoms**

1. The bacterium attacks all the plant parts above ground level at different stages of plant growth.
2. The earliest symptom of the disease is seen in the cotyledons of geminating seeds. Minute, water-soaked spots appear on the under-surface of the cotyledons. These later increase in diameter, tum brown to black and form irregular patches distorting the shape of the cotyledons causing them to dry and wither.

Fig. 45: Bacterial blight in cotton leaves

3. The disease spreads to new leaves formed and the seedling may ultimately collapse and die. On the leaves similar water-soaked spots appear on the under surface first and then on the upper surface. They increase in size, become angular, bound by small veinlets of the leaf and tum brown to black. Often the disease spreads along the edge of the veins, hence called Vein Blight or Black Vein. Sometimes large areas are formed due to coalescence of a number of small spots leading to death and shedding of leaves.
4. The infection may also spread to petioles causing them to collapse. In the affected areas large amounts of

bacterial slime are exuded which form a dry film on the brown lesions.

5. Lesions on stem, petioles, and fruiting branches are dark brown to sooty black. They are elongated and sunken. The affected stems show cracks and gummosis and are easily broken by wind or there may be girdling and death of affected organs. These are the black arm symptoms.

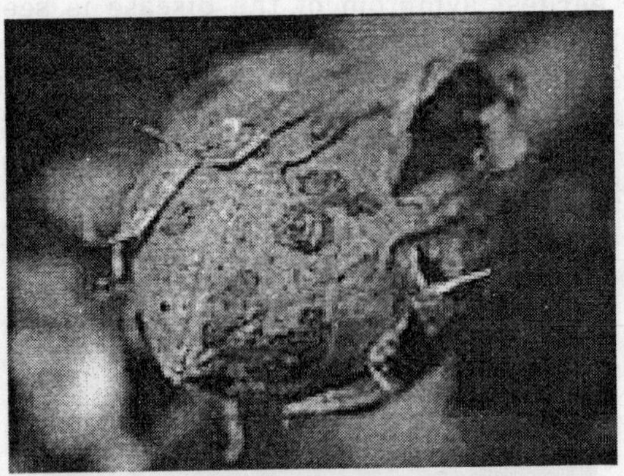

Fig. 46: Bacterial blight in cotton boll

6. On the bolls or fruits the disease is characterized by the appearance of water-soaked lesions on the surface. These lesions turn dark brown and finally black, and are invariably sunken. Yong infected bolls fall down prematurely. If they mature, lint is of not much commercial value. The bacterium within the boll passes along the fibres and infects the seed externally. It may also reach the interior of the seed either through micropyle or through punctures.

(iii) Control Measures

1. Removal and destruction of diseased plant debris are recommended to reduce the soil-borne inoculum. Deep ploughing after harvest buries the infected stalks and thus reduces survival ability of the bacterium in soil. Presowing irrigation to enable the left over seeds germinate, followed by poughing and then planting of the main crop can be followed.

2. Destruction of possible alternate or collateral hosts is also essential. Crop rotation, late sowing, early thinning, good tillage, early irrigation and addition of potash to soil help in reducing the disease incidence.

3. This can be eliminated by seed treatments. External inoculum on the seed is destroyed by delinting of seed with concentrated sulphuric acid. Seeds are immersed in acid for 10-15 minutes, then rinsed thoroughly by suspending in water to remove acid, and finally dried and treated with organomercurial compounds like Agrosan GN, Ceresan, etc. However, this treatment does not destroy internally seed-borne inoculum. Treating the seed with antibiotics like streptomycin eradicates the internally seed-borne infection. The bacterium is known to live in the seed for about a year or so and, therefore, ageing of the seed for two years before sowing has also been recommended. Hot water treatment of seed at 56°C for 10 minutes destroys the external and internal inoculum without affecting seed viability

4. The secondary spread can be checked by regular spraying with copper fungicides (0.2 to 0.3 per cent). First spray is given when the crop is five to six weeks old. In all three to six sprayings, depending on the severity of the disease, are given at 15 days interval. Mathur, et al. (1973) have reported that seed dressing of cotton with Agrimycin (3 gm/40 kg) and its spray (25 ppm) are most promising in controlling the black arm of cotton.

5. Control through development of resistant varieties is possible and provides the best and most effective preventive measure. There have reported the following nine cultivars of hirsutum group as resistant: 70 IH-480/2, 70 IH-480/3, 70 IH-480/9, K-4005, Badnawar-1, B1007, Khandwa-2 DHY-286, M-937-CTO-421.

2. Name of disease: Citrus Canker

(i) Causal Organism: *Xanthomonas campestris* pv *citri* (Hasse) Dye or
Xanthomonas citri (Hasse) Dowson.

(ii) Symptoms

1. The disease occurs on leaves, twigs, thorns, older branches and fruits. Leaf lesions first appear as small, round, watery, translucent spots. They are raised and become yellowish brown.
2. They first develop on the lower surface of the leaf and then on both the surfaces.
3. As the disease advances the surface of the spots becomes white or greyish and finally ruptures in the centre giving a rough, corky, and crater-like appearance. The spots increase in size (1 mm to 1 cm in diameter), and may coalesce to form elongated lesions on fruits and twigs. The rough lesions are surrounded by a yellowish-brown to green raised margin and watery yellow halo.

Fig. 47: Bacterial canker in *Citrus* leaves.

4. Spots occurring of petioles and midrib cause premature defoliation.
5. On larger branches the necrotic lesions (cankers) are irregular, rougher, and more prominent.

Fig. 48: Bacterial canker in *Citrus* fruit

6. Cankers on fruits are similar to those on leaves except that the yellow halo is absent and a crater-like depression in the centre is more prominent. The injury to fruits is only skin deep and no effect on the pulp or juice is noticed.
7. Cankers on twigs cause them to break.

(iii) Control Measures

1. The disease is so serious that the only practical method of control is complete destruction of the diseased plants by burning them. Though costly, this method is most effective.
2. Use of disease-free nursery stodk for planting in new orchards.
3. Spray the plants before planting in new orchards with 1 per cent Bordeaux mixture.
4. In old orchards pruning of the affected twigs and spraying with 1 per cent Bordeaux mixture at periodical intervals, especially during rainy season.
5. The dropped off canker-affected leaves and twigs should be collected and burnt.
6. The vigour of the plant should always be maintained by proper irrigation and fertilization

7. Proper care should be taken to minimize the attack of leaf miners. Quarantine regulations prohibiting movement of diseased stock should be rigidly followed.
8. Disease can be controlled by antibiotic sprays. Streptomycin sulphate or crude agricultural preparations of streptomycin at 500 to 1000 ppm concentration sprayed at 15-day interval effectively checked the disease on 48-year old lime trees while Bordeaux mixture was ineffective. Phytomycin (2500 ppm) also was effective. The antibiotics are absorbed by leaves and translocated thus functioning as a systemic bactericide.
9. Spray of neem-cake at the rate of about 160 Ibs per acre is highly effective in checking citrus canker as well as leaf miner. Fifty pounds of cake is soaked in 20 gallons of water and allowed to decompose for about a week. It is then sprayed without filtration. Some of the cake falls on the ground and becomes manure. Several sprayings are required to produce good results.
10. Control through resistant varieties is also possible as different species of *Citrus* show different degrees of susceptibility to the disease.

3. Name of Disease: Bacterial Leaf Blight of Rice

(i) Causal Organism: *Xanthomonas campestris* pv. *oryzae* (Ishiyama) Dye or

Xanthomonas oryzae, (Ishiyama) Dowson.

(ii) Symptom

1. The symptoms of the disease vary considerably with the stage of infection and the prevailing weather conditions. Leaf blight phase is more commonly seen. The blight phase is characterized by linear yellow to straw coloured stripes with wavy margin, generally on both edges of the leaf, rarely on one edge. These stripes usually start from the tip and extend downwards. This is followed by the drying and twisting of the leaf tip and rapid extension of marginal blight lengthwise and crosswise to cover larger areas of the leaf.

Fig. 49: Bacterial leaf blight of rice

2. In occasional cases the linear stripes may develop anywhere on the lamina or along the midrib with or without the marginal stripes. The blighting may extend to the leaf sheaths and culms, killing the tiller or the whole clump.
3. In dry weather opaque and turbid drops of bacterial ooze which dry into yellowish beads can be seen on the leaf surface. These drops are washed down by rains `
4. The glumes of seeds also get infected but the symptoms are not well defined.
5. The blight phase of the disease usually appears four to six weeks after transplanting. Since leaf drying can be due to physiological disorders or attack of tungro virus also, the diseased leaves should be subjected to microscopic examination for correct diagnosis. If the affected portions of leaves are cut and mounted in a drop of water on a glass slide bacterial ooze can be seen as a cloudy mass at the cut ends.
6. The most destructive form of the disease in the tropics is 'Kresek' or wilt phase resulting from early systemic infection or from infected seed and the bacterium brought in contact with germinating seedlings.
7. The leaves roll competely, droop, turn yellow or grey and ultimately the tillers wither away. In severe cases the

affected stool may be completely killed. The Kresek affected tillers can be confused with stem borer injury but the latter can be easily pulled out while it is not so with. Kresek affected tillers.

8. Another common symptom found in the topics is the pale yellow leaf phase. Some of the youngest leaves in a clump turn pale yellow or whitish. These leaves later turn yellowish brown and wither away.

(iii) Control Measures

1. It has observed that 95 per cent eradication of infection in seeds by soaking for 12 hours in 0.025 per cent water solution of Agrimycin (an antibiotic containing 15 per cent streptomycin and 1.5 per cent oxytetracycline) plus 0.05 per cent wettable Ceresan and then transferring the seed to hot water at 52° to 54°C for 30 minutes.
2. In another experiment, it has observed that simply soaking of seed in water for 12 hours followed by exposure to hot water at 53°C for 30 minutes is enough to eradicate the seedborne inoculum.
3. It has observed that dipping the seed for eight hours in 0.1 per cent Ceresan wet plus Streptocyclin at 0.3 g in 10 litres of water had a significant effect in controlling the initial infection. The spread of secondary infection can be checked to a great extent by spraying with this antibiotic (3 g/100 litres).
4. The antibiotic Streptocyclin has been most commonly recommended chemical for seed treatment and foliar spray against bacterial blight of rice in India.
5. Variety N-22 is highly resistant to the disease. Among newer cultivarrs IR-20, IR-22 and Ratnahave mild resistance. These varieties can be used to control the plant disease.

4. Name of Disease: Bacterial Leaf Streak of Rice

(i) Causal Organism: *Xanthomonas campestris* pv. *oryzicola* (Fang, et al.) Dye or

Xanthomonas translucens f. sp. *oryzicola* (Fang, et al.) Bradbury

(ii) Symptoms

1. Bacterial leaf streak is a foliar disease. The first sign of the disease is the appearance of fine, watersoaked to translucent interveinal streaks which may be as long as 1 to 10 cm. These streaks are restricted by the veins and soon turn yellow or orange brown.

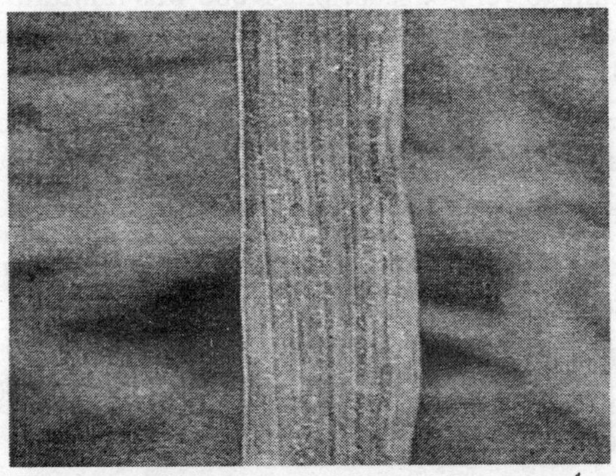

Fig. 50: Bacterial leaf streak of rice

2. Minute yellow or amber beads of bacterial exudates are abundant on the streaks. When these beads dry, streaks of rough pustules may be felt on the leaf. These streaks may coalesce to form large patches and cover the entire leaf surface.
3. Eventually, the leaves may be completely blighted.
4. In highly susceptible varieties streaks are surrounded by yellow halo.
5. The infection may reach the leaf sheath and even the seed coat but symptoms are not very clear.

(iii) Control Measures

1. The seed must be obtained from a reliable source to minimize the danger from seed-borne inoculum.
2. Seed soaked in 0.025 per cent streptocycline and hot water treatment at 52°C for 30 minutes are effective in eradicating the seed infection.

3. It has been repòrted that the spraying of Vitavax at 0.15 to 0.3 per cent to be effective in preventing infection and lesion development.
4. Sankel, Captan and Fytolan were also effective to some extent.
5. 1,118 varieties screened by artificial inoculation found resistant to very susceptible. None was immune and only 140 varieties showed few and small lesions indicating resistance.
6. In India IR-20, Krishna and Jagannath have shown good tolerance to this disease.

5. Name of Disease: Bacterial Spot of Tomato

(i) Causal Organism: *Xanthomonas campestris* pv. *vesicatoria*

(ii) Symptoms

1. Bacterial spot in tomato is more noticeable on frruit but not seriously injure leaves.

Fig. 51: Bacterial leaf spot of tomato

2. Infect green fruits have slightly raised spots that are to inch in diameter and br with rough surfaces.
3. Ripe fruits are not infected.
4. This disease infect plants in the seedling stage then spread to green fruits takes place during weather.

(iii) Control Measures

1. Copper containing fungicides like Burgundy mixture, chaubattia paste etc. help hold this spread in check.

6. Name of Disease: Black Rot of Crucifers

(i) Causal Organism: *Xanthomonas campestris* pv. *campestris* (Pammel) Dowson

(ii) Symptoms

1. This disease is found on a number of cruciferous vegetables like *Beta valgaris, Brassica oleracea* var. *botrytis, Cucumis sativus* etc.

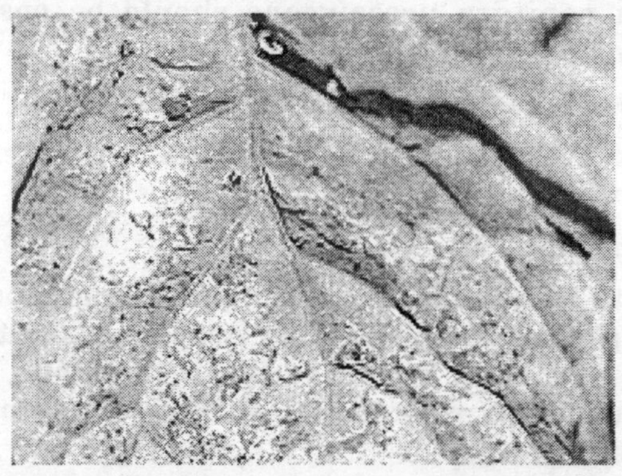

Fig. 52: Black rot of crucifers

2. The first sign of the disease appears near the leaf margin, characterized by chlorosis, which progresses towards the centre of the leaf blade.
3. In the affected portions, the veins and veinlets turn brown and finally black.
4. In case of severe infection, the leaves wither.
5. The vascular blackening may extend to the main stem, causing systemic spread of the infection.
6. From the affected plant parts, bacterial ooze may also be seen.

(iii) Control Measures

1. Seed treatment with 0.1% mercuric chloride for half an hour is effective in eradicating seed-borne infection.

2. Hot water treatment of the seed at 50°C for 30 minutes is also effective.
3. Drenching the seed bed with 5% commercial formalin may help in checking the disease.
4. In a small scale field trial, Agrimycin 100 gave good control of the disease.
5. Streptocycline was found to be effective in checking the "*in vitro*" growth of pathogen.
6. The crop should be rotated. A three-year rotation is desirable.
7. Strict sanitary precautions should be observed to minimize the incidence of black rot.

7. Name of Disease: Leaf Spot of Mango

(i) Causal Organism: *Xanthomonas campestris* pv. *mangiferaeindici* (Patel, et al.; Robbs, et al.).

(ii) Symptoms

1. Minute water-soaked lesions appear in groups towards the tip of the leaf blade. They increase in size to about 1 to 4 mm, turn brown to black in colour, and are surrounded by chlorotic halos. They are surounded by the veins.
2. Large necrotic patches may be formed by coalescing of several lesions. The patches sometimes dry up. These patches are often rough and raised due to heavy bacterial exudate.
3. When a greater portion of the lamina surface has been affected the leaves may fall.
4. Petioles, fruits, and tender stems may also be infected. On young fruits water-soaked lesions appear and turn dark brown to black. Cracks may appear in the skin of the fruit. Badly affected fruits drop prematurely.

(iii) Control Measures: No specific control measures have been developed against this disease.

8. Name of Disease: Bacterial Blight of Beans

(i) Causal Organism: *Xanthomonas campestris* pv. *phaseoli*

(ii) Symptoms

1. Common bacterial blight of beans disease is favored by higher temperatures (greater than 80°C) usually prevalent during July and August. Symptoms usually appear on older leaves, even at low humidity.

Fig. 53: Bacterial blight of beans

2. It exhibits a scalded appearance on leaf tissue and contain water-soaked spots. These small, irregular shaped lesions often enlarge to greater than one inch or more and form dark brown lesions along the edge of the leaflet.
3. A narrow lemon-yellow margin often surrounds these lesions and large portions of the foliage can be infected.
4. Infected pods exhibit circular, water-soaked areas that often produce yellow massess of bacterial ooze.
5. Later, spots dry and appear as reddish-brown lesions. Pod infection often causes discolouration, shriveling and bacterial contamination of seeds; however, some seed appear healthy.

(iii) Control Measures

1. Overwintering populations can effectively be reduced by employing a three-year rotation with beans planted once every third year. Corn, small grains and vegetables are recommended crop alternatives during two of the three years.

2. Proper sanitation of bean crop debris is important. It requires the complete incorporation of straw into the soil and the elimination of volunteer beans early the following year. This reduces diseased plants which provide the inoculum responsible for disease outbreaks in nearby bean fields. Bacterial pathogens can also survive in bean dust on contaminated harvest equipment, seed-cleaning equipment and storage containers.

3. Always plant high quality certified seed to minimize early season disease. Treat seeds with streptomycin, which effectively kills most external bacterial contamination. Diseases can still develop in these fields, but the pathogens must be transported into the field from external inoculum sources.

4. Avoid cultivation of beans when plants are wet or when the stand is too tall to allow machinery to pass through without wounding the foliage. Thoroughly clean soil, weeds and crop residues from equipment before moving to other fields. Reuse of irrigation runoff water is not recommended since pathogens are transmitted by it. Sprinkler irrigation can increase the spread and severity of bacterial foliar diseases by splashing bacteria from plant to plant, and extending leaf wetness periods.

5. Planting resistant varieties is the best method to manage bacterial diseases of beans. Most older bean varieties, especially pintos and light red kidneys, are susceptible to these bacterial diseases. Fortunately, blight resistant varieties are available in other market classes.

6. Common bacterial blight is not controlled as effectively by copper spray programs, although some reduction in disease is obtained with a preventive program. Applying these protectants early in the season every seven to 10 days during cool moist weather can decrease establishment of bacterial pathogens.

9. Name of Disease: Bacterial Blight of Ornamental Plants

(i) Causal Organisms: *Xanthomonas campestris* pv. *pelargoni*

(ii) Names of Affected Ornamental Plants

1. *Geranium*
2. *Dieffenbachia*
3. *Philodendron*
4. *Syngonium*
5. *Aglaonema*

(iii) Symptoms

1. The invasion by the bacterium leads to very rapid and extensive necrosis of the affected plant parts resulting in scorched appearance of the lost.
2. The bacterial exudates on leaves provide secondary inoculum which is spread by rain splash, water or insects.

(iv) Control Measures

1. Purchase culture-indexed plants known to be free of the most important bacterial pathogens
2. Discard infected plants.
3. Do not use overhead irrigation.
4. Pasteurize the propagation bed and medium between crops.
5. Do not handle soil or debris on the potting soil surface and then the plant.
6. The disease can be controlled by antibiotics and chemical spray.

10. Name of Disease: Bacterial Leaf Spot of Foliage Plants

(i) Causal Organism

1. *Xanthomonas campestris* pv. *dieffenbachiae*
2. *Xanthomonas campestris* pv. *hederae*

(ii) Names of Affected Foliage Plants

1. *Philodendron* spp.
2. *Dieffenbachia* spp.
3. *Anthurium* spp.
4. *Hedera helix*

(iii) Symptoms

1. This disease is most active under hot, humid conditions.
2. The most common symptom is yellowing along leaf margin beginning at the leaf tip.
3. Under hot, humid conditions, the leaf margin may turn reddish brown rather than yellow.
4. Early symptoms of infection are small, translucent dots which then turn yellow.
5. The centers of older lesions often turn brown.
6. As the disease progresses, affected leaves turn yellow and drought from the stem.
7. In English Ivy (*Hedera helix*), leaf spots are light green and translucent with a red margin, older spots turn brown or black.
8. Leaf stalks become black and shriveled. This decay may extend down to twigs and woody stems, and definite cankers may be seen.

(iv) Control Measures

1. Avoid placing plants where there are conditions of high humidity, crowding, or poor air circulation. Do not mist plants and avoid wetting the foliage when watering, as bacteria need water to multiply and spread to healthy leaves. Water plants according to recommendations, being careful not to overwater them. Proper watering, repotting every 6 months to 1 year in fresh sterile soil, fertilizing every 8-12 weeks during the spring and summer, and controlling insect infestations will keep plants growing in healthy condition and lessen the likelihood of infestation by bacteria.
2. Provide conditions that are optimum for the plant's growth. Isolate the diseased plant and prune infected leaves, but avoid excessive handling of diseased plants. If more than one third of the plant is involved, prune infected leaves over a period of time, since removing too many leaves at one time will put the plant under further stress. Disinfect scissors before each cut by dipping them

into a freshly made solution of 1 part Chlorox or Hilex bleach and 9 parts water.

11. Name of Disease: Bacterial Leaf Blight of Foliage Plants

(i) Causal Organism : *Xanthomonas* spp.

(ii) Names of Affected Foliage Plants :

1. *Syngonium* spp.
2. *Aglaonema roebellinii*
3. *Aglaonema* spp.

(iii) Symptoms

1. Syngonium spp. are most often attacked by this bacterium
2. Symptoms include translucent lesions at the leaf tip and along the leaf margin.
3. The lesions may elongate and extend into the middle of the leaf. Lesions are dark green at first, then turn yellow and eventually turn brown when dead.
4. The diseased area is often surrounded by a bright yellow halo that separates it from the healthy portion of the leaf.
5. White flakes of dried bacterial exudate are often generally on older lesions on the undersides of leaves.

(iv) Control Measures

1. Avoid placing plants where there are conditions of high humidity, crowding, or poor air circulation. Do not mist plants and avoid wetting the foliage when watering, as bacteria need water to multiply and spread to healthy leaves. Water plants according to recommendations, being careful not to overwater them. Proper watering, repotting every 6 months to 1 year in fresh sterile soil, fertilizing every 8-12 weeks during the spring and summer, and controlling insect infestations will keep plants growing in healthy condition and lessen the likelihood of infestation by bacteria.
2. Provide conditions that are optimum for the plant's growth. Isolate the diseased plant and prune infected

leaves, but avoid excessive handling of diseased plants. If more than one third of the plant is involved, prune infected leaves over a period of time, since removing too many leaves at one time will put the plant

II. Pseudomonas Species and Their Plant Diseases

They are most serious bacterial plant pathogens of the world. Their cells are single; rods are straight or curved put not helical; generally 0.5-1.0X1.5X4 μ in size. They have one or more polar flagella. They are gram negative, strictly aerobes, chemoorganotroph and non-spore forming. Their metabolism is respiratory and they never be fermentative. *Pseudomonas* have total 87 species; of which about 24 are plant pathogenic. There are some important bacterial diseases caused by *Pseudomonas* are described below:

1. Name of Disease: Bacterial Speck of Tomato

(i) Causal Organism: *Pseudomonas syringae* pv. *tomato*

(ii) Symptoms

1. This disease is found on leaves and is most noticeable fruits.
2. Only young, green fruits are susceptible.
3. Severe losses occasionally occur, especial early crops.
4. This disease often infects plants in the seedling stage; spreads to green fruits takes place during weather.

Fig. 54: Bacterial speck of tomato

(iii) Control Measures

1. Copper containing fungicides like Burgundy mixture, Chaubattia paste etc. help hold this spread in check.

2. Name of Disease: Bacterial Brown Rot of Potatoes or Wilt Disease of Potatoes

(i) Causal Organism: *Pseudomonas solanacearum* (Smith) Smith.

(ii) Symptoms

1. This disease is not only found in potato but also in other solanaceous crops like chillies, tomato, eggplants. Apart from these crops it is also found in a large number of other cultivated and wild plants including castor, groundnut, banana and ginger.
2. The characteristic symptoms of the disease are wilting, stunting, and yellow refers to the browning of the xylem in the vascular bundles. This browning is often visible from the surface of the infected stems as dark patches or streaks.
3. The name ring disease is derived from the fact that a brown ring is formed in the tuber due to discolouration of the vascular bundles. The skin of infected tubers is often discoloured.
4. In severely affected tubers the eye buds are blackened. If the infected stems or tubers are cut across and squeezed, grayish-white bacterial ooze comes out of the vascular ring. Microscopic examination of properly stained slides of the ooze confirms the presence of the disease.
5. In many cases, when nematode infection is also present, the stem at the base of tomato plants becomes dark brown and constricted, leading to toppling down of the plant.
6. During continued humid weather when the temperature is also high the most conspicuous symptoms of the wilt on tomato plants due to this disease have been observed to be sudden drooping of leaves and rotting of the stem

from any point. The roots appear healthy and often are well developed.

(iii) Control Measures

1. Since the disease is mainly soil borne, proper rotation helps in its control. Three year rotation with maize, soybean and red top grass (*Agrostis alba*) has been found to afford considerable protection to the crop. In India, two-year rotation with potato-wheat-sanai green manuring-wheat-green manure-potato has been suggested. It has been reported that rotations consisting of cowpea-maize-cabbage or okra-cowpea-maize are good for reducing tomato wilt and maize-okra-radish, maize-cowpea-maize or okra-cowpea-maize reduce eggplant wilt caused by the bacterium
2. Since the disease is a systemic vascular wilt and the pathogen perpetuates in diverse soil types, its control through chemicals has not been possible. Application of high dosages of nitrogen (500 kg urea per acre). toxic chemicals like sulphur, chloropicrin, etc., which are effective cannot be recommended because of the high cost. Under experimental conditions foliar application of Agrimycin 100, chloramphenicol or streptomycin sulphate a day prior to inoculation are effective at 1000 ppm but not after inoculation
3. Differences in susceptibility to brown rot occur in varieties of *Solanum tuberosum* but so far no resistant cultivars have been developed. *S. phureja* has shown resistance in some of its clones and its inclusion in hybridization programme for inducing resistance in *S. tuberosum* is being treid.
4. Always obtain seeds from areas where the disease is not reported.
5. In endemic areas the seed tubers should be selected only from healthy crop and should be treated in 0.02 per cent streptocycline for 30 minutes after giving 5 mm deep cut.
6. The plots which show severe incidence of brown rot should be put under maize, cereals or soybean for three years.

7. Rain or irrigation water should not be allowed to flow from diseased fields to healthy plo ts.
8. Infected plant residue should always be removed and destroyed.

3. Name of Disease: Red Stripe of Sugarcane

(i) Causal Organism: *Pseudomonas rubrilineans* (Lee, et al.) Stapp.

(ii) Symptoms

1. The disease first appears as water-soaked elongated streaks which soon become chlorotic and carry dark-red stripes which are 0.5 to 1 mm in width and 5 to 100 mm or more in length
2. Sometimes two or more stripes coalesce to form larger bands.
3. The lower half of the leaf is more affected than the tip.
4. When young shoots are affected symptoms of top rot appear.
5. The growing point of the shoot shows many dark-red stripes water-soaked appearance and undergoes rotting.
6. The disease proceeds downwards killing the terminal buds and the leaves.

(iii) Control Measures

1. When once the disease sets in, it is difficult to control it, Systematic cutting down and burning of the affected shoots may help in reducing the spread.
2. Use of healthy setts for seed is recommended. Growing resistant varieties seems to be the best method to avoid this disease.

4. Name of Disease: Wild fire of Tobacco

(i) Causal Organism: *Pseudomonas tobaci*

(ii) Symptoms

1. Plants of all age are attacked but seedlings are affected more. The first symptoms appear on the leaves of poorly growing seddlings which show an advancing wet rot at

the margins and tips with a water soaked zone separating the rotting and healthy tissues.

2. The whole leaf or only parts of it may rot and fall off.
3. Seedlings may be killed in the seed bed.
4. The most common symptoms are round, yellowish-green spots on leaves which turn brown within a day. The brown spots are surrounded by yellowish green haloes.
5. The brown spots and chlorotic haloes enlarge and coalesce to form large, irregular, dead areas.
6. In wet conditions the leaves fall off and give a form appearance.

(iii) Control Measures

1. The seed borne infection of this disease can be prevented by soaking the seeds in formaldehyde solution for 10 minutes.
2. Spray of neutral copper fungicide and streptomycin also control the disease.
3. Tobacco varieties rasistant to wild fire bacteria are also available.

5. Name of Disease: Halo Blight of Beans

(i) Causal Organism: *Pseudomonas syringae* pv. *phaseolicola*

(ii) Symptoms

1. Halo blight symptoms on leaves initially appear as greasy, water-soaked spots about 1/16 inch in diameter.
2. They are generally most visible on the underside of young leaflets approximately one week after infection. One or more water-soaked spots are later surrounded by a greenish-yellow halo about 1/16 to 1/2 inch in diameter. This halo is caused by a bacterial toxin.
3. The size and development of this halo is variable, and may not be produced during high temperatures.
4. During severe infection, the disease may become systemic and cause yellowing and death of new foliage.

Fig. 55: Halo blight of bean leaves

5. Halo blight symptoms on pods appear as small circular, water-soaked spots or streaks. A light cream or silver-coloured bacterial ooze is often associated with these spots. Pod infection often causes discoloration, shriveling and bacterial contamination of seeds; however, some seed may appear to be healthy.

(iii) Control Measures

1. To ensure complete coverage, these products should be applied with at least 5 gallons of water per acre. When pathogens have become established and disease symptoms are evident, copper sprays can help reduce their spread to healthy foliage and pods. But bacteria inside lesions are not affected. Commonly used bactericides include Kocide, NuCop and Champ. Always read the product label carefully and consult extension and industry representatives for updated recommendations.
2. Most navy and small white varieties, including Aurora, are resistant to halo blight. Great Northern varieties such as Beryl, Harris, Ivory and Marquis are resistant to halo and common bacterial blights. The red kidneys Foxfire and Chinook are resistant to halo blight.
3. Closely inspect bean fields for symptoms of halo blight and bacterial brown spot throughout the growing season,

especially after prolonged periods of high humidity. Apply a bactericide every seven to 10 days if infection is detected and weather conditions remain favorable for disease development. Hail-damaged beans should receive a preventive application of a copper bactericide to protect damaged plants during regrowth.

6. **Name of Disease: Brown Spot of Beans**

(i) **Causal Organism:** *Pseudomonas syringae* pv. *Syringae*

(ii) **Symptoms**

1. Bacterial brown spot symptoms are similar to young halo blight lesions after initial infection.
2. The symptoms appear as small water-soaked spots most visible on the underside of young foliage.

Fig. 56: Brown spot of bean pods

3. A narrow, greenish yellow border about 1/16 inch wide may surround some of the lesions. However, when the lesion matures, it typically develops a "brown spot" appearance, and dead tissue in the center may fall out, producing a shot-hole appearance.
4. Infected pods may be twisted and kinked and exhibit circular brownish water-soaked spots.

(iii) **Control Measures**

1. Avoid placing plants where there are conditions of high humidity, crowding, or poor air circulation. Do not mist

plants and avoid wetting the foliage when watering, as bacteria need water to multiply and spread to healthy leaves. Water plants according to recommendations, being careful not to overwater them. Proper watering, repotting every 6 months to 1 year in fresh sterile soil, fertilizing every 8-12 weeks during the spring and summer, and controlling insect infestations will keep plants growing in healthy condition and lessen the likelihood of infestation by bacteria.

2. Provide conditions that are optimum for the plant's growth. Isolate the diseased plant and prune infected leaves, but avoid excessive handling of diseased plants. If more than one third of the plant is involved, prune infected leaves over a period of time, since removing too many leaves at one time will put the plant under further stress. Disinfect scissors before each cut by dipping them into a freshly made solution of 1 part Chlorox or Hilex bleach and 9 parts water.

7. Name of Disease: Leaf Spot of Dracaena Sanderana

(i) Causal Organism: *Pseudomonas* spp

(ii) Symptoms

1. A leaf spot produces translucent lesions, sometimes with thin, reddish-brown margins.
2. Leaves turn yellow, and affected areas become dry and papery.

(iii) Control Measures: Same as above.

8 Name of Disease: Leaf Spot of Ornamental Plants

(i) Causal Organism: *Pseudomonas cichorii*

(ii) Symptoms

1. leaf spots on chrysanthemum, geranium, impatiens, and many other ornamental plants.
2. The spots are generally water-soaked (wet-looking) and dark brown to black. Depending upon the plant infected, the leaf spots may have a yellow halo.

(iii) Control Measures: Same as above.

9. Name of Disease: Bacterial Leaf Spot of Foliage Plants

(i) Causal Organism: *Pseudomonas cichorii*

(ii) Susceptible Plants

(a) *Epipremnum aureum* (Pothos),

(b) *Philodendron panduraeforme* (Fiddleleaf Philodendron),

(c) *Aglaonema* spp. (Chinese Evergreen), and

(d) *Monstera* spp. (Split-leaf Philodendron.)

(iii) Symptoms

1. They are varied and may include brownish-black lesions, light and dark zones on Epipremnum aureum leaves, and a yellow halo around affected areas on Monstera deliciosa leaves.

(iv) Control Measures: Same as above.

III. Erwinia Species and Their Plant Diseases

Their cells are predominantly single. Rods are straight and their size is measured approximately 0.5-1.0X1.0-3.0μ. They have peritrichous flagella for motility except two species i.e. *Erwinia stewartii* and *Erwinia dissolvens*. They are gram-negative and grow well on artificial media aerobically as well as anaerobically and produce acid. The genus has three major groups of species, one group consisting of pectolytic bacteria that cause soft rot (often called pectobacterium species) and the second group consisting of those that do not cause soft rot but dry necrosis and wilt, etc. A third group consists of epiphytes that cause neither soft rot nor necrosis. The genus is divided into species on the basis of plant pathogenesis and biochemical-physiological differences. This genus has total 20 species. There are some important bacterial diseases caused by *Erwinia* are described below.

1. Name of Disease: Bacterial Stalk Rot of Maize

(i) Causal Organism: *Erwinia chrysanthemi* pv. *zeae* (Sabet) Victoria, et al.

(ii) Symptoms

1. The disease occurs on young as well as old plants. The rot occurs at the lower nodes and passes up and down the stalk to a very limited extent.

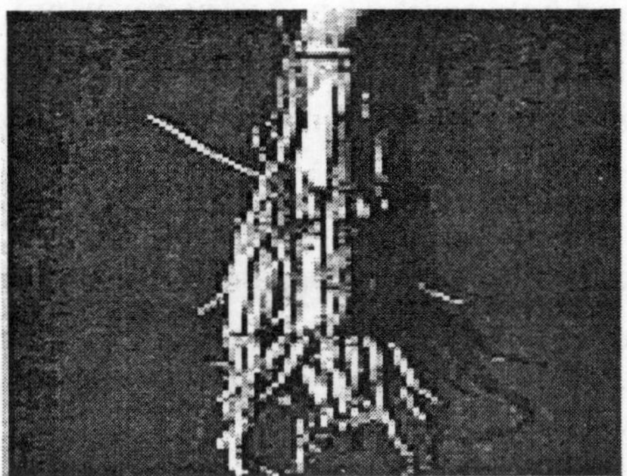

Fig. 57: Bacterial stalk rot of maize

2. At times the rot involves the rind only, causing a dark soft decay there, while at other times the rotted area is principally in the interior of the stalk.
3. With the advance of stalk rot, leaves may start yellowing and drying.
4. The infected tissues of the stalk are first soft but later on they turn into a dry mass of shredded, easily disjointed fibres. At this stage the plant topples down.
5. In addition to stalk rot, a light to dark brown rottting of the sheath, starting from the point where it is attached to the affected tissues of the stalk has also been seen during the rainy season.

(iii) Control Measures

1. No definite control measures have been recommended for this disease. As early as 1965, Thompson (1965) had demonstrated the efficacy of chlorinated water in the sprinkler irrigation system in the control of bacterial stalk rot of maize.
2. Two applications of Klorocin containing 22 per cent chlorine as soil drench at the rate of 25 kg/ha, first before flowering and second 10 days later gives good control.
3. Chlorine persists for 10 days in roots and basal stem portions and for 15 days in the upper parts.
4. Use of insecticides to ward off insect attack can also be recommended.

5. Plant and field sanitation is very important.
6. Accumulation of water around stalks should be avoided.
7. Stubbles of the diseased crop should be collected and burnt.
8. Ganga 2 is the least susceptible variety.

2. Name of Disease: Black Leg and Soft Rot of Potato

(i) Causal Organism

(a) *Erwinia carotovora* subspecies *carotovora* (Jones) Bergey, et al.

(b) *Erwinia carotovora* subspecies *atroseptica* (Van Hall) Dye.

(ii) Symptoms

1. In the field the typical 'black-leg' is characterized by a striking brown black or jet black colour of the stem at the soil level.
2. This discolouration usually starts from the old seed tuber.
3. The cortical tissues may shrivel and rot.
4. The plants may achieve normal height but usually they remain dwarfed and stunted.
5. Instead of spreading normally, the branches and leaves show a tendency to grow upwards.
6. The foliage turns light green or yellow, with a slight metallic luster, and soon wilts and dies.
7. Curling of leaves similar to that caused by potato leaf curl virus may also be found.
8. Sometimes young seedlings arising from diseased seed tubers are destroyed before or soon after emergence.
9. In soft rot of tubers, which may occur in the field if the soil is moist and temperature is high, or during transit and storage, the tubers are transformed partly or totally, slowly or quickly, into a soft decayed pulpy mass. This mass is held together only by the corky epidermis which cannot be attacked by the parasite.

10. When a soft rot tuber is cut open the colourless putrid mass turns a pink red on exposure to air, rapidly becoming brownish-red to brown black. The cut tubers have only a slight smell. However, when secondary bacteria have gained entry into the rotting mass a strong repulsive smell is felt.

(iii) Control Measures

1. Dipping the whole tubers in fungicidal or bacterial solutions is of no use as the bacterium is deep seated. When cut tubers are used for planting, the best thing to do is to keep the cut tubers at low temperature (12° to 15°C) for about four days. This allows cork formation to take place and avoid entry of soil-borne bacteria. Alternatively, the cut tubers should be dipped in mercuric chloride solution, organo-mercury compounds (Agallol or Aretan) or hot formalin solution.
2. Use only healthy tubers as seed material.
3. Avoid using cut potatoes as seed.
4. Plant less deeply in heavy soil than in light soil.
5. Do not plant too early or too late.
6. Give best tillage to the field but avoid injuring young plants.
7. As soon as a diseased plant is located it should be dug out and burnt.
8. Avoid injury to tubers during harvesting, transit and storage.
9. Remove plant debris and tuber material from the field after harvest and destroy them.
10. Pick out the diseased and injured tubers before packing or storing the tubers.
11. Wash the tubers with chlorinated water before storage.
12. Keep the stores dry, well ventilated, and cool.
13. If possible there should be periodical check of the store and picking and destruction of diseased tubers.

3. Name of Disease: Bacterial Wilt of Cucurbits

(i) Causal Organism: *Erwinia tracheiphila* (Smith) Bergey, et al.

(ii) Symptoms

1. It affects the plant by causing sudden wilting of foliage and vines, and finally death of the plant.
2. It also causes slime rot of squash fruit in storage.
3. On the cut surface of infected stems, droplets of bacterial ooze appear.

Fig. 58: Bacterial wilt of cucurbit

(iii) Control Measures

1. The disease is controlled best by controlling the cucumber beetles with insecticides like, carbaryl (sevin), methoxychlor, rotenine, etc.
2. Antibiotics such as streptomycin, terramycin and neomycin at high concentration (500 ppm) give good control of the disease.

4. Name of Disease: Fire Blight of Apple and Pears

(i) Causal Organism: *Erwinia amylovora* (Burrill) Winslow

(ii) Symptoms

1. The first symptoms of fire blight appear usually on the flowers which become water soaked, then shrivel rapidly, turn brownish to black in colour and sometimes fall off.

2. Soon the symptoms spread to the leaves on the same spur or on nearby twigs, starting as brown-black blotches along the midrib and main veins or along the margins and between the veins.
3. As the blackening progresses, the leaves curl and shrivel, hang downwards and usually cling to the curled, blighted twigs.
4. Terminal twigs and water sprouts (suckers) are usually infected directly and wilt from the tip downwards.
5. Their bark turns brownish-black and is soft at first but later shrinks and hardens.
6. The tip of the twig is hooked and the leaves turn black and cling to the twing
7. From fruit spurs and terminals, the symptoms progress down to the supporting branches where they form cankers.
8. The bark of the branch around the infected twig appears water soaked at first, later becoming darker, sunken and dry.
9. If the canker enlarges and encircles the branch, the part of the branch above the infection dies.
10. If the infection stops short of girdling the branch, it becomes a dormant or inactive canker, with sunken and sometimes cracked margins.
11. Infection usually takes place through the pedicel, but direct infection is not uncommon. The fruit becomes water-soaked, turns brown, shrivels, mummifies and finally turns black. Dead fruits may also cling to the tree for several months after infection.
12. Under humid conditions, droplets of a milky-coloured, sticky ooze may appear on the surface of any recently infected part. The ooze usually turns brown soon after exposure to the air. The droplets may coalesce to form large drops which may run off and form a layer on parts of the plant surface.

(iii) Control Measures

1. Several measures need to be adopted to successfully control fire blight. Control measures include excision and

destruction of overwintering cankers or twig infections before the blossoms open.

2. The application of 0.1% mercuric chloride solution was found to be effective in checking primary infections.
3. One or two sprays of 1% bordeaux mixture have been moderately successful.
4. Application of streptomycin at 100 ppm three times during the pre-blossom and blossom period effectively controlled the disease.
5. A special insect trap has been developed by which the bees emerging from the hives pass through the streptomycine dust which covers them and thus gets transmitted to the flowers. In the flower the antibiotic gets distributed and iinhibits the growth of the bacterium.

5. Name of Disease: Stem Rot of Ornamental Plants

Erwinia chrysanthemi and *Erwinia carotovora* survive in plant debris that is not completely decomposed, on or in infected plants, on other greenhouse plants without causing disease, and under some conditions, in soil. Both species infect a wide range of plants in the greenhouse. *E. chrysanthemi* has been shown to survive on plants that it does not actually infect. They can cause a stem rot or mushy, brown, smelly, soft rot.

6. Name of Disease: Bacterial Blight of Foliage Plants

(i) Causal Organism: *Erwinia chrysanthemi*

(ii) Name of Affected Plants

(a) *Aglaonema* spp. (Chinese Evergreen),

(b) *Dieffenbachia* spp., Philodendron spp.,

(c) *Syngonium* spp.

(iii) Symptoms

1. The bacteria-attack some plants systemically (internally), especially Dieffenbachia spp. Symptoms of systemic infection are the yellowing of new leaves, wilting, and a mushy, foul-smelling stem rot.

2. Aerial spread of this bacterium can cause foliar infection. Symptoms may appear as rapid, mushy leaf collapse on Philodendron spp., definite leaf spots on Syngonium spp., or all of these symptoms on Philodendron selloum.
3. Erwinia chrysanthemi grows best in warm-to-hot, wet, and humid environments. Attack by these bacteria often results in the death of foliage plants.

(iv) Control Measures

There is no specific measure for controlling this disease.

III. Agrobacterium Species and Their Plant Diseases

The cell shape of *Agrobacterium* is found in L-form or L-phase which lack rigid cell wall. They are gram-negative, and rod shaped. They have peritrichous flagellation for motility. Their colonies mostly white, do not hydrolyse starch. They cause hapertrophy of affected plant tissues. They can grow on relatively simple medium but its isolation from the soil requires selective media. They consist of a heterogenous group of stains. There is an important bacterial disease caused by *Agrobacterium* is described below:

1. Name of Disease: Crowngall of Stone Fruits

(i) Causal Organism: *Agrobacterium tumefaciens* (E.F. Smith and Towns) Conn.

(ii) Names of Affected Stonefruits

(a)	Apple	*(f)*	Rose
(b)	Peach	*(g)*	Pear
(c)	Apricot	*(h)*	Cherry
(d)	Almond	*(i)*	Plum
(e)	Grape		

(iii) Symptoms

1. Small outgrowths appear on the stem and roots. When young, the galls are soft, spherical, white or flesh-coloured.
2. The galls vary in size from 7 mm to 100 mm in diameter.

3. On woody stems, the galls are hard and corky. They are generally knobby and knotty and become more cleft as they grow older.
4. The affected plants may become stunted with chlorotic leaves.
5. This is primarily a disease of the parenchyma.
6. Infection results in the rapid proliferation of cells.
7. Unlike other galls, the crown gall is an example of non self-limiting tumour.

(iv) Control Measures

1. Crop sanitation is necessary to avoid the introduction of infected material into the nursery stock.
2. Antibiotics such as Vancomycin, Aureomycin, Streptomycin and Terramycin have been tried to prevent this disease.

V. Clavibacter Species and Their Plant Diseases

The genus was proposed by Davis, et al. (1984) for those phytopathogenic *Corynebacterium* spp,. that contain diaminobutyric acid as the major amino acid in their cell-wall and which were formerly included in the *Corynebacterium michiganense* group such as *C. michiganense, C. nebraskense, C. insidiosum, C. rathayi, C. sepedonicum, C. iranicum* and *C. tritici*. These workers also reported the identification of the sugarcane ratoon stunting disease bacterium as *Clavibacter xyli* subsp. *xyli*.

The cells are gram-positive, non-acid fast, pleomorphic rods, often arranged at an angle to give V-formation as a result of snapping or bending type of cell division; no coccoid cells are seen; non-endospore forming; non-motile; strict aerobes; nutritionally exacting; nitrate not reduced; cell-wall peptidoglycan contains diaminobutyric acid; G+C content of the DNA is 70± 5 moles per cent. Type species is *Clavibacter michiganense* (*Corynebacterium michiganense*).

The following species of *Corynebacterium* have been transferred to genus *Clavibacter*:

1. *Corynebacterium michiganese* (Smith) Jensen or *C. michiganense* pv. *michiganense* (Smith) Dye and Kemp,

or *C. michiganense* subspecies *michiganense* (Smith) Carlson and Vidaver renamed as *Clavibacter michiganense* subspecies *michiganense* (Smith) Davis, et al. It causes canker of tomato and chilli.

2. *Corynebacterium sepedonicum* (Spieckemann and Kotthoff) Skaptason and Burkholder or *C. michiganense* pv. *sepedonicm* or *C. michiganense* subspecies *sepedonicum* renamed as *Clavibacter michiganense* subspecies *sepedonicum* (Spieckemann and Kitthoff) Davis, et al. It causes wilt and ring rot of potato.

3. *Corynebacterium nebraskense* (Schuster, et al.) Vidaver and Mandel or *C. michiganense* subspecies *nebraskense* (Schuster, et al.) Carlson and Vidaver renamend as *Clavibacter michiganense* subspecies *nebraskense* (Schuster, et al.) Davis, et al. It causes wilt and blight of maize.

4. *Corynebacterium insidiosum* (Mc Culloch) Jenson or *C. michiganense* subspecies *insidiosum* (Mc Culloch) Carlson and Vidaver renamed as *Clavibacter michiganense* subspecies *insidiosum* (Mc Culloch) Davis, et al. It causes wilt of alfalfa.

5. *Corynebacterium tritici* (Hutchinson) Burkholder or *C. tritici* (ex Hutchinson) Carlson and Vidaver renamed as *Clavibacter tritici* (Carlson and Vidaver) Davis, et al. It causes yellow ear rot or tundly disease of wheat.

6. *Corynebacterium rathayi* (Smith) Dowson or C. michiganense pv. rathayi (Smith) Dye and Kemp renamed as *Clavibacter rathayi* (Smith) Davis, et al. It causes a gumming disease of some graminaceous plants.

7. *Corynebacterium iranicum* Scharif or *C. michiganense* pv. *iranicum* (Scharif) Dye and Kemp or *C. iranicum* (ex Scharif) Carlson and Vidaver renamed as *Clavibacter iranicum* (Carlson and Vidaver) Davis, et al. It causes a wheat disease similar to yellow ear rot.

8. *Clavibacter xyli* Davis, et al., 1984 is a new species. *C. xyli* subspecies *xyli* causes ratoon stunting disease of sugarcane and *C. xyli* subspecies *cynodontis* causes stunting disease of Bermuda grass.

There are some important bacterial diseases coused by Clavibactor are described below.

1. Name of Disease: Bacterial Ear Rot of Wheat

(i) Causal organisms: *Clavibacter tritici* (Carlson and Vidaver, 1982; Davis et al., 1984).

(ii) Symptoms

1. Wrinkling of the lower, and twisting of the central, leaves is the first symptom of the disease
2. This is accompanied by exudation of a bright yellow sticky slime enveloping the entire ear.
3. This slime binds the glumes, stem, and leaf sheath thus checking the plant growth and distorting the stem.
4. In wet weather the slime starts trickling down but in dry weather it becomes a deeper yellow, hard and dry.
5. The disease appears only when the crop is reaching maturity.

(iii) Control Measures

1. Sowing of wheat seeds free from nematode galls is an essential measure for control of this disease. The galls can be removed by floating the seed in brine (40 lbs salt in 25 gallons of water).
2. Seeds should always be obtained from places where disease is not occurring.
3. Since the galls may survive in soil thereby prepetuating the bacterium, attempts should be made to destroy them.
4. Affected plants should be located, uprooted carefully, and burnt. The field should be well drained.

2. Name of Disease: Ratoon Stunting Disease of Sugarcane

(i) Causal Organism: *Clavibacter xyli* subspeciess *xyli* sp. nov.

(ii) Symptoms

1. The name 'ratoon stunting' is a misnomer since the disease equally affects plant as well as ratoon crops.

Diseased stools usually show stunted growth, reduced tillering. thin canes with shortened internodes and yellowish leaves.

2. Such abnormalities in growht are not specific for this disease and could be due to many otherr factors such as lack of water and nutrients. If such symptoms appear in the field one has to first exclude the possibility of poor growth due to mismanagement. If the non-pathogenic causes are not existing. RSD can be suspected.

3. Geminability of diseased setts is low. Germination is delayed and plant growth is slow. Thus, weeds usually become more predominant and this not only further hampers crop growth but also escalates cultivation cost. During summer diseased plants show signs of wilting.

4. Typical symptoms of RSD are seen only after splitting open the cane longitudinally. Two types of discolouration can be seen in the pith. In mature canes, there is orange red discolouration of vascular bundles at the nodes. In young canes pink colour can be seen near the nodes. However, many varieties are symptomless carriers and the presence of RSD can be detected only by using indicator sugarcane varieties or other hots.

(iii) Control Measures

Since the pathogen is neither soil borne nor true seed borne and there is no spread in the standing crop, it is obvious that if setts are disease free the disease can be controlled. The following steps are recommended for managment of the disease:

1. Any good yielding variety can be grown in a separate plot only for seed. The plot should have good drainage and optimum dosages of fertilizers and irrigation should be ensured. The crop should also receive timely hoeing and weeding operations. All weeds growing on the borders should be destroyed. If weak plants are seen in the crop they should be uprooted and burnt. At the same time roguing of stools showing symptoms of wilt, red rot

and smut should also be done. Special attention should be given to the even germination of planted setts. Any sett that fails to germinate within a reasonable period should be dug out and removed.

2. Seed setts should be taken from only those fields or parts of the field where the plants are robust. Weak crop should never be used for seed.
3. Weak or stunted crop should not be ratooned.
4. Seed setts may be given hot water, hot air, or aerated steam therapy, wherever feasible. In India, the time-temperature relation recommended for hot water treatment is 50°C for two hours or 52°C for half hour. Precautions necessary in this treatment are use of small wire net baskets to facilitate proper and uniform exposure of setts to hot water which should be constant at 50° or 52°C for duration of the treatment, cooling of setts immediately after treatment, and selection of seed material for treatment from a field where there is a low incidence of the disease. The hot air therapy involves dry treatment at 54°C for eight hours. Bud damage is less in hot air than in hot water therapy. The aerated steam therapy or moist heat therapy is considered better.

Although antibiotics are known against coryneform bacteria, chemotherapy of RSD has not been found successful so far. No RSD immune variety of sugarcane is grown comercially anywhere in the world.

3. Name of Disease: Bacterial Canker in Tomato

(i) Causal Organism: *Clavibacter michiganense*

(ii) Symptoms

1. Initial symptoms of bacterial canker, caused by Clavibacter michiganense, are wilting and curling of leaves on part of the plant.
2. Later these leaves turn brown but do not fall. Disease symptoms spread throughout the plant in a few days or over several weeks. Plants may die prematurely.

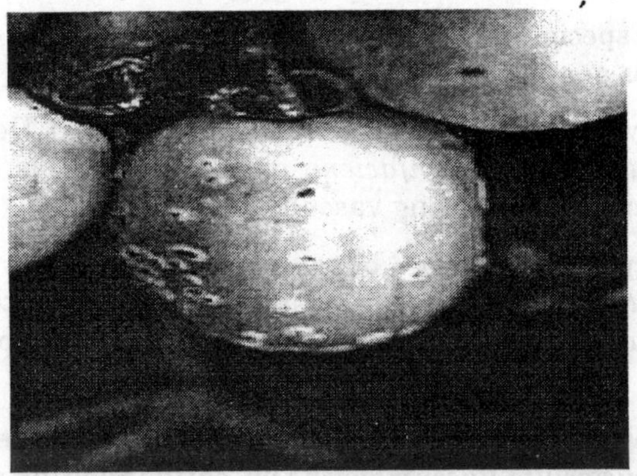

Fig. 59: Bacterial canker in tomato

3. Other disease symptoms include longitudinal stem cracks (2-3 inches) and small (inch) fruit spots with persistent white halo-like margins.

(iii) Contro Measure

To control this disease, destroy infected plants and obtain transplants from a different source the following year.

VI. Curtobacterium Species and Their Plant Disease

Prior to 1980, the plant pathogenic corynebacteria (Gram-positive, pleomorphic, non-branching rod- shaped bacteria) were assigned to the genus *Corynebacterium* although they did not exactly satisfy the description of a typical *Corynebacterium*. After 1980, these plant pathogenic species were transferred to other genera such as *Curtobacterium*, *Clavibacter*, *Rhodococcus* and *Arthrobacter* on the basis of typical chemistry of cell-wall, G+C content of the DNA and genetic homologies.

The genus *Curtobacterium* Yamada and Komagata has the following characters: Cells small short rods, coccoid cells found in old cultures; weekly gram-postive, frequently old cells lose gram-positivity; generally motile by lateral flagella; cell multiplication by bending type of cell division; pleomorphism only slight; major cell-wall amino acid is ornithine; G+C content of the DNA ranges from 66 to 71 moles per cent.

The species of *Corynebacterium* that have been classified under this genus are:

Corynebacterium flaccumfaciens (Hedges) Dowson as *Curtobacterium flaccumfaciens* pv. *flaccumfaciens* (Hedges) Collins and Jones causing vascular wilt of bean.

Corynebacterium betae Keyworth, et al., as *Curtobacterium flaccumfaciens* pv. betae (Keyworth, et al.) Collins and Jones causing vascular wilt and leaf spot of red beet.

Corynebacterium oortii Saaltink and Mass Gaesteranus as *Curtobacterium flaccumfaciens* pv. *oortii* (Saaltink and Mass Gaesteranus) Collins and Jones causing vascular wilt and leaf and bulb spot of tulip.

Corynebacterium poinsettiae (Starr and Pirone) Burkholder as *Curtobacterium flaccumfaciens* pv. *poinsettiae* (Starr and Pirone) Collins and Jones causing stem canker and leaf spot of poinsettia.

There is an important disease caused by *Curtobacterium* is described below:

1. Name of Disease: Bacterial Wilt of Beans

(i) Causal Organism: *Curtobacterium flaccumfaciens* ssp. flaccumfaciens).

This disease is reported in some regions but the symptoms are not available in literature.

VII. Rhodococcus Species and Their Plant Diseases

This genus belongs to family Nocardiaceae of Actinomycetales. *Corynebacterium facians* (Tilford) Dowson causing a fasciation disease of many plant species is the only plant pathogen reclassified under this genus as *Rhodococcus fascians* (Tilford) Goodfellow. cells in young culture measure 0.5-0.9 × 1.5-4.0 microns, are slightly curved rods arranged singly, at angles or as pallisade; Gram-positive and non-motile; nitrate not reduced; cell-wall peptidoglycan contains diaminopimelic acid as main amino acid; G+C content of the DNA is 62.9 to 67.3 moles per cent.

There is an important disease caused by *Rhodococcus* is described below:

1. Name of Disease: Abnormal Branching of Ornamental Plants

(i) Causal Organism: *Rhodococcus fascians* (formerly *Corynebacterium fascians*)

(ii) Symptoms

1. Abnormal branching and stem development near the base of infected plants such as geranium.
2. The bacterium is carried on infected cuttings and may enter the propagation medium.

(iv) Control Measures: Same as above.

VIII. Streptomyces Species and Their Plant Diseases

This is the best recognized genus of the family Streptomycetaceae (Actinomycetales) and is widely distributed in nature, especially in soil. The genus as described in the Bergey's Manual (1974) has the following characters:

Slender, coenocytic filaments, 0.5-2.0 microns in diameter; aerial mycelium at maturity forms chains of three to many spores; cell-walls contain diaminopimelic acid; gram-positive; aerobic; heterotrophs; generally reduce nitrates. On isolation, colonies are small, 1 to 10 mm in diiameter, descrete and lichenoid, leathery or butyrous, initially relatively smooth but later develop a weft of aerial mycelium that may appear granular, powdery, velvety or floccose.

The Bergey's Manual had listed 463 species in the genus but only 275 were retained in the Approved lists of 1980. Only two species, *Streptomyces ipomea* and *S. scabies* are recognized as plant pathogens although many others have been isolated from diseased plant specimens.

There is an important plant disease caused by *Streptomycin* in described below:

1. Name of Disease: Common Scab of Potato

(i) Causal Organism: *Streptomyces scabies* (Thaxter) Waksman and Henrici

(ii) Symptoms

1. There are two types of lesions formed on the tuber: (i) the shallow and (ii) the deep scab
2. In shallow scab the affected tubers show superficial roughened areas, sometimes raised above. and often slightly below the plane of the healthy skin. The lesions consist of corky tissue which arise from abnormal proliferation of the cells of the tuber periderm due to attack of the pathogen. The lesions vary widely in size and shape, only sometimes darker than the healthy skin, and often become confluent to give a reticulate appearance.
3. In deep pitted scab the lesions measure 1 to 3 mm or more in depth and are darker than the lesions in shallow scab. They also are corky and may join together so that the entire tuber surface becomes affected. The deep pitted lesions are either extensions of the shallow lesions, combined effect of the scab organism and some chewing insects, or due to some specific strain of the scab organism.

(iii) Control Measures

1. Due to its soil and seed borne nature the control has been very difficult in almost all the countries. Since soil pH and soil moisture are important in the development of this disease, these have been explored for minimizing the losses. In addition, attempts have been made to induce biological control through such practices as green manuring and soybean cover crop.
2. Tubers with deep lesions should not be used for seed. As far as possible select only blemish free seed tubers. The tubers can be disinfected by dipping them for five minutes in 0.25 per cent suspension of mercurial fungicides such as Emisan-6 Agallol-6 or Aretan. This treatment has been recommended for control of black scurf disease of potato also.
3. Supression of activity of the pathogen in the soil through modification of soil environments (pH, moisture, and

management of soil microbiota) has been extensively studied. Application of elemental sulphur to reduce pH to below 5.2 has been successful but because of high cost involved it is not a practical measure.

4. Scab is controlled field plots by maintaining soil moisture near field capacity during tuber formation, until all the internodes of the tubers were formed. Thus, maintaining the soil moisture at field capacity from the fifth week after planting up to the ninth week by frequent light irrigation is recommended.
5. Green manuring and cover crops of soybean promote activity of antagonistic bacteria. Therefore, in infested fields green manuring before planting or cultivation of legumes before potato may be uselful. A rotation of four years avoiding beet, fleshy rooted crucifers and carrot, which are susceptible to this pathogen, is also beneficial.
6. All actinomycetes are susceptible to pentachloronitrobenzene. Thus, application of PCNB (Brassicol) at the rate of 20 to 30 kg/ha will reduce not only common scab but also the black scurf or Rhizoctonia disease of potato.

IX. *Ralstonia Solanacearum* (formerly *Pseudomonas solanacearum*) and Their Plant Diseases

It causes vascular wilting of many herbaceous ornamentals, including geraniums. Gross symptoms in geraniums mimic those caused by *Xanthomonas campestris* pv. *pelargonii*. Unlike most other bacteria, *Ralstonia solanacearum* survives well in the soil. Once a greenhouse is contaminated with this organism, it is difficult to eliminate and poses a threat to many different crops. Symptoms include wilting, discolouration of the vascular tissue, leaf yellowing and death of the plant.

References

Allen T., Shen P., Samsel L., Liu R., Lindahl L., and Zengel J.M. 1999. Phylogenetic analysis of L4-mediated autogenous control of the S10 ribosomal protein operon. J Bacteriol 181:6124-6132.

Anil Kumar T.B. and B.P. Chakrabarti. 1970. Factors affecting survival of Erwinia carotovora, causal organism of stalk rot of maize, in soil. Acta Phytopath. 5:333-340.

Avery O.T., C.M. MacLeod and M. Mccarty 1944. Studies on the chemical nature of the substance inducing transformation of pneumococcal types. Induction of transformation by a desoxyribonucleic acid fraction isolated from pneumococcus type-III. J. Exp. Med. 79: 137.

Bhide V.P. 1948. A comparative study of some wilt producing phytopathogenic bacteria. Indian Phytopath. 1:70 79.

Bhide V.P.l949. Utilization of organic nitrogen by Corynebacterium michiganense (E.F.S.) Dowson and its possible bearing on parasitism. Indian Phytopath 2: 173-175.

Bhide V.P., R.K. Hegde and M.K. Desai 1956. Bacterial red stripe disease of sugarcane caused by Xanthomonas rubrilineans in Bombay state. Curr. Sci. 25:30.

Bocchetta M., Gribaldo S., Sanangelantoni A. and Cammarano P. 2000. Phylogenetic depth of the bacterial genera Aquifex and Thermotoga inferred from analysis of ribosomal protein, elongation factor and RNA polymerase subunit sequences. J Mol Evol, 50:366-380.

Borneman,J . and Triplett E.W. 1997. Molecular microbial diversity in soils from eastern Amazonia: evidence for unusual microorganisms and microbial population shifts associated with deforestation. Appl Environ Microbiol, 63:2647-2653.

Bost F., Diarra-Mehrpour M. and Matin J.P. 1998. Inter-a-trypsin inhibitor proteoglycan family. A group of proteins binding and stabilizing the extracellular matrix. Eur J Biochem 252:339-346.

Boyle A.M. and R.M. Price 1969. Vancomycin prevents crown gall. Phytopath., 53: 1272-1275.

Buchanan R.E. and N.E. Gibbons 1974. Bergey's Manual of determinative bacteriology 8th Ed. Williams and Wilkins, Baltimore.

Buddenhagen, I. W. 1964. Bacterial wilt of certain seed bearing Musa sp. caused by tomato strain of Pseudomonas solanacerum. Phytopathology 54: 286.

Burkholer H.W. 1948. Bacteria as plant pathogens. Ann. Rev. Microbiol. 2: 389-412.

Cairns J. 1963 The bacterial chromosome and its manner of replicating as seen by autoradiography. J Mol. Bioi. 6: 208.

Carlson R.A. and A.K. Vidaver 1982. Taxonomy of Corynbacterium plant pathogens including a new pathogen of wheat, based on polyacrylamide gel electrophoresis of cellular proteins. Int. J. Syst.Bacteriol. 32: 315-326.

Chakravarti B.P. and M. Rangarajan 1967. A virulent strain of Xanthomonas oryzae isolated from rice seeds in India. Phytopath.57: 688-690.

Chakravarti N.K. and S. Devadath 1971. Identification of source of resistance to bacterial blight of rice. (Abs), second Inter. Symp. pl. Pathol. IARI, New Delhi, p.22.

Chang P and Marians K.J. 2000. Identification of a region of Escherichia coli DnaB required for functional interaction with DnaG at the replication fork. J Biol Chem 275:26187-26195.

Chattopadhyay S.B. .and N. Mukherjee 1968. Occurrence in nature of collateral hosts (Cyperus rotundus and C. defformis) of Xanthomonas oryzae, incitant of bacterial blight of rice., Curr. Sci., 37: 441-442.

Chauhan M.S., M.S. Kairon and S.S. Karwasra 1983. Determination of minimum number of effective sprays of Agrimycin plus Blitox for the control of bacacrial blight of cotton under Haryana conditions Indian Mycol. Plant Pathol. 13: 187-191.

Cheema S.S., S.P. Kapur and R.D. Bansal. 1985. Efficacy of various therapeutic agents against greening disease of citrus. Journal of Researchl, PAU. 22: 479-482.

Chenuil A., Soliganc M., and Bernard M. 1997. Evolution of the large-subunit ribosomal RNA binding site for protein L23/25. Mol Biol Evol, 14:578-588.

Chopra B.L., T.H. Singh and J.R. Sharma 1974. Reaction of different varieties of cotton to angular leaf spot (Xanthomonas malvacearum).IndianJ. Mycol. Plant Path. 4: 85-86

Choudhury H. 1935. A bacterial disease of wheat in Punjab. Proc. Ind. Acad. Sci., 1(B): 579-585

Chu S., Noonan B., Cavaignac S., and Trust T.J .1995. Endogenous mutagenesis by an insertion sequence element identifies Aeromonas salmonicida AbcA as an ATP-binding cassette transport protein required for biogenesis of smooth lipopolysaccharide. Proc Natl Acad Sci USA, 92:5754-5758

Cook A.A., J.C. Walker and R.H. Larson 1952. Studies on the disease cycle of black rot pathogen of crucifers. Phytopath. 42: 162-167.

Damann K.E. Jr, K.S. Derrick A.G. Gillaspie Jr., D.B. Fontenut and J. Kao. 1978. Detection of the RSD bacterium by serologically specific electron microscopy. Proc. 16th Congr. Intern Soc.Sugarcane Technol. pp. 433-437.

Davis M.J., A.G. Gillaspie Jr., A.K. Vidaver and R.W. Harris. 1984. Clavibacter: a new genus containing some phytopathogenic coryneform bacteria including Clavibacter xyli subspecies xyli and C.xyli subspeciescynodontis, pathogens that cause ratoon stunting disease.Int.. J. Syst. Bacteriol.34:107-117.

Davis M.J., A.G. Gillaspie Jr., R.W. Harris and R.H. Lawson 1980. Ratoon stunting disease of sugarcane: Isolation of the causal bacterium. Science 210: 1365-1367.

DeLong E.F., Franks D.G. and Alldredge A.L. 1993. Phylogenetic diversity of aggregate-attached vs. free-living marine bacterial assemblages. Limnol Oceanogr 1, 38:924-934

Derakshani M., Lukow T. and Liesack W. 2001. Novel bacterial lineages at the (sub)division level as detected by signature nucleotide-targeted recovery of 16S rRNA genes from bulk soil and rice roots of flooded rice microcosms. Appl Environ Microbiol, 67:623-631.

Derbyshire V., Grindley N.D. and Joyce C.M .1991. The 3'-5' exonuclease of DNA polymerase I of Escherichia coli: contribution of each amino acid of the active site to the reaction. EMBO J 10:17-24.

Devadath S. 1971 (Abstract), Second Intern. Symp. PI. Pathol., IARI, New Delhi, pp. 139-141.

Devadath S. and S. Y. Padmanabhan 1969. Approaches to the control of bacterial blight and streak diseases of rice in India.lndianPhytopahl. 23:153-154.

Devadath S. and S.Y. Padmanabhan. 1971. Factors influencing the incidence of baderial blight of rice and its control. Oryza, Suppl. 8:381-392.

Doidge E.M. 1915. A bacterial disease of mango. Ann. Appl. Bioi. 1: 1-45.

Drlica K. 1987. The nucleoid. p. 91-103. In F. C. Neidhardt, J. L. Ingraham, K. B. Low, B. Magasanik, M. Schaechter, and H. E. Umbarger (ed.), Escherichia coli and Salmonella typhimurium: cellular and molecular biology. American Society for Microbiology, Washington, D.C.

Durgapal J.C. 1971. A preliminary note on bacterial diseases of temperate plants in India-II. Bactcrial disease of stone fruits. Ind, Phytopath, 24: 379-382.

Dye D. W., J.F. Bradbury, M. Goto, A.C. Hayward, R.A. Lelliott and M.N. Schroth. 1980. International standards for naming pathovars of phytopathogenic bacteria and a list of pathovar names and pathotypestrains. Rev. Plant Pathol, 59: 153-168.

Embley T.M., Hirt R.P. and Williams D.M. 1994. Biodiversity at the molecular level: the domains, kingdoms and phyla of life. Phil Trans R Soc Lond B 345:21-33.

Fuerst J.A .1995. The planctomycetes: emerging models for microbial ecology, evolution and cell biology. Microbiology 141:1493-1506.

Garnier M. and J.M. Bove. 1983. Transmission of the organism associated with citrus greening disease from sweet orange to periwinkle by dodder. Phrytopathology 73: 1358-1368.

Gillaspie A.G., R.E. Davis and J.F. Worley 1976. Nature of the bacterium associated with ratoon stunting disease of sugarcane. Sugarcane Pathol. Newsletter 15/16: 11-15.

Giovannoni S.J., Godchaux W., Schabtach E. and Castenholtz R.W. 1987. Cell wall and lipid composition of Isosphaera pallida, a budding eubacterium from hot springs. J Bacteriol, 169:2702-2707.

Goto M. 1965. Resistance of rice varieties and species of wild rice to bacterial leaf blight and bacterial leaf streak diseases. Philippine Agr. 48: 329-338.

Graham D.C. 1963. Serological diagnosis of potato black-leg and tuber soft rot. Plant Pathology 12:142-144.

Grifiith F. 1928. Significance of pneumococcal types. J. Hug. Camb. 27:113.

Guillouet S., Rodal A.A., An G., Lessard P.A. and Sinskey A.J .1999. Expression of the Escherichia coli catabolic threonine dehydratase in Corynebacterium glutamicum and its effect on isoleucine production. Appl Env Microbiol, 65:3100-3107.

Guo D., Bowden M.G., Pershad R. and Kaplan H.B. 1996. The Myxococcus xanthus rfbABC operon encodes an ATP-binding cassette transporter for O-antigen biosynthesis and multicellular development. J Bacteriol, 178:1631-1639.

Gupta R. 2000. The phylogeny of proteobacteria: relationships to other eubacterial phyla and eukaryotes. FEMS Micro Rev 24:367-402.

Hayes W. 1968. The genetics of bacteria and their viruses, 2nd ed., Blackwell, Oxford.

Hingorani M.K., P.P. Mehta and N.J. Singh 1956. Bacterial brown rot of potatoes in India. Indian Phytopath, 9: 67-71.

Hingorani M.K. and S.K. Addy 1952. Comparative study of Erwinia carotovora, E. aroideae and E. atroseptica. Indian Phytopath. 5: 40-43.

Hingorani M.K. and S.K. Addy 1953. Factors influencing bacterial rot of potato. Indian Phytopath., 6: 110-115.

Hingorani M.K., P.P. Mehta and N.J. Singh 1956. Bacterial brown rot of potatoes in India. Indian Phrytopath. 9: 67-71.

Hirel P.H., J.M. Schmitter, P. Dessen, G. Fayat and S. Blanquet 1989. Extent of N-terminal methionine excision from Escherichia coli proteins is governed by the side-chain length of the penultimate amino acid. Proc. Natl. Acad. Sci. USA 86:8247-8251.

Hoch J.A.1993. Regulation of the phosphorelay and the initiation of sporulation in Bacillus subtilis. Annu. Rev. Microbiol. 47:441-465.

Hollis J.P. and R.W. Goss. 1950. Factors influencing invasion of potato by Erwinia carotovora. Phrytopathology 40: 860-869.

Hugenholtz P., Pitulle, C. Herschberger, K.L. and Pace N.R. 1998. Novel division level bacterial diversity in a Yellowstone hot spring. J Bacteriol, 180:366-376.

Hunter R.E. and L.A. Brinkerhoff 1963. Internally infected seeds as a source of inoculum for the primary cycle of bacterial blight of cotton. Phytopath., 53: 1397-1401.

Hward A.C. 1964. Characteristics of Pseudomonas solanacearumJ. Appl. Bact.27: 265-277.

Ishiyam S. 1922. Studies on bacterial leaf blight. Rep. Agr. Exp. Sta, 45: 233-261.

Jacob F. and E.L. Wollman 1962. Sexuality and the genetics of bacteria. Academic Press, New York.

Jain S.S., P. Rangareddy and S.Y. Padmanabhan 1966. Control of leaf blight of rice caused by Xanthomonas oryzae (Uyeda and Ishiyama) Dawson. Indian Phytopath. 19:29.

Jenkins C., Kedar V. and Fuerst J.A. 2002. Gene discovery within the planctomycete division of the domain Bacteria using sequence tags from genomic DNA libraries. Genome Biol 3(6) RESEARCH 0031.

Jenkins C., Kedar V., Fuerst J.A. 2002. Gene discovery within the planctomycete division of the domain Bacteria using sequence tags from genomic DNA libraries Genome Biol: (6): RESEARCH0031.

Jenkins C. and Fuerst J.A. 2001. Phylogenetic analysis of evolutionary relationships of the planctomycete division of the domain Bacteria based on amino acid sequences of elongation factor-Tu. J Mol Evol, 52:405-418.

Johnson R. C., C. A. Ball, D. Pfeffer and M. I. Simon 1988. Isolation of the gene encoding the Hin recombinational enhancer binding protein. Proc. Natl. Acad. Sci. USA 85:3484-3488.

Kaiser D. and R. Losick 1993. How and Why Bacteria Talk to Each Other. Cell73:873-885.

Kapur S.P. and S.S. Cheema 1983. Chemotherapeutic control of citrus greening disease. Pesticides 17: 13.

Kelman A. and L. Sequeira 1965. Root to root spread of Pseudomomonas solanacearum Phytopathology 55: 1618-1619.

Kim C.H. and Y.S. Cho 1970. Effects of N.P.K. fertilizer levels and growth condition on the development of bacterial leaf blight of rice plant. J. Pl. Protec., Korea, 9: 7-13.

Kirby J.R., Niewold T.B., Maloy S. and Ordal G.W. 2000. CheB is required for behavioural responses to negative stimuli during chemotaxis in Bacillus subtilis. Mol Microbiol 35:44-57.

Kojima K., S. Iwase and K. Inoki 1955. Relationship between the soil type and the occurrence of bacterial leaf blight disease of rice. Ann. Phytopath. Soc. Japan 19:63.

Koonin E.V., Mushegaian A.R. and Rudd K.E. 1996. Sequencing and analysis of bacterial genomes. Curr Biol, 6:404-416.

Krishna A. and A.G. Nema 1983. Evaluation of chemicals for the control of citrus canker. Indian Phytopath. 36:348-349.

Kulkarni N.B. 1962. A note on disease of paddy in Maharashtra. Dhuia. Agr. Coll. Mag., 1:65-77.

Lal S. and S.C. Saxena 1982. Appraisal of yield loss due to bacterial stalk rot of maize. Indian J. Mycol. Plant Pathol. 12:308-309.

Lal S. and S.C. Saxena 1982. Field evaluation of calcium hypochlorite for the control of bacterial stalk rot of maize. Indian J. Mycol. Plant Pathol. 12:278-282.

Lal S., B.S. Thind, M.M. Payak and B.L. Renfro 1970. Bacterial stalk rot of maize. Resistance breeding and chemical control. Bull. Indian Phytopath. Soc; 31:24-27.

Lapage S.P., P.H.A. Sneath E.F. Lessel V.B.D. Skerman H.P.R. Seelinger and W.A. Clark 1975 International code of nomenclature of bacteria, Amer. Soc. Microbiol., Washington.

Leary B.A., Ward-Rainey N. and Hoover T.M. 1998. Cloning and characterization of Planctomyces limnophilus rpoN: complementation of a Salmonella typhimurium rpoN mutant strain. Gene, 221:151-157.

Liesack W. and Stackebrandt E. 1992. Occurrence of novel groups of the domain Bacteria as revealed by analysis of genetic material isolated from an Australian terrestrial environment. J Bacteriol, 174:5072-5078.

Liesack W., König H., Schlesner H. and Hirsch P. 1986. Chemical composition of the peptidoglycan-free cell envelopes of budding bacteria of the Pirella/Planctomyces group. Arch Microbiol, 145:361-366.

Lindsay M.R., Webb R.I. and Fuerst J.A.1997. Pirellulosomes a new type of membrane-bounded cell compartment in planctomycete bacteria of the genus Pirellula. Microbiology 143:739-748.

Lindsay M.R., Webb R.I., Strous M., Jetten M.S., Butler M.K., Forde R.J. and Fuerst J.A. 2001. Cell compartmentalisation in planctomycetes: novel types of structural organisation for the bacterial cell. Arch Microbiol, 175:413-429.

Losick R. and P. Stragier 1992 . Crisscross regulation of cell-type-specific gene expression during development in Bacillus subtili. Nature 355:601-604.

Martin C. and U. Nydegger 1982. Susceptibility of Cyphomandra betacea to Pseudomonas solanacearum. Plant Disease 66: 1025-1027.

Mathur R.L., L.N. Daftari and S.L. Jhamaria 1973. Controlling black arm of cotton by fungicides and antibiotics.Indian J. Mycol. Plant Pathol. 3:107-108.

May A.P. and Ponting C.P.1999. Integrin a- and ß4 subunit-domain homologues in cyanobacterial proteins. Trends Biochem Sci 24:12-13.

Mehrotra R. S. 1988. Plant Pathology Fifth Edition, Tata McGraw-Hill Publishing Company Ltd. New Delhi.

Midha S.K. and Gopal Swarup 1972. Factors affecting development of ear cockle and tundu diseases of wheat. Indian J . Nematol. 2: 97-104.

Minamino T. and Macnab R.M. 1999. Components of the Salmonella flagellar export apparatus and classification of export substrates. J Bacteriol 181: 1388-1394.

Miyamoto S., Teramoto H., Coso O.A., Gutkind J.S., Burbelo P.D., Akiyama S.K. and Yamada K.M. 1995. Integrin function: molecular hierarchies of cytoskeletal and signalling molecules. J Cell Biol, 131:791-805.

Mohanty S.K., P.R. Reddy and R. Sridhar 1982. Effect of calcium and magnesium on susceptibility of rice plants to bacterial leaf blight Curr. Sci. 51: 298-299.

Moolenaar G.F., Moorman C. and Goosen N. 2000. Role of the Escherichia coli nucleotide excision repair proteins in DNA replication. J Bacteriol. 182:5706-5714.

Murphy L. D. and S. B. Zimmerman 1997. Isolation and characterization of spermidine nucleoids from Escherichia coli. J. Struct. Biol. 119:321-335.

Murphy L. D. and S. B. Zimmerman 1997. Stabilization of compact spermidine nucleoids from Escherichia coli under crowded conditions: implications for in vivo nucleoid structure. J. Struct. Biol. 119:336-346.

Neef A., Amann R., Schlesner H. and Schleifer K.H. 1998. Monitoring a widespread bacterial group: in situ detection of planctomycetes with 16S rRNA-targeted probes. Microbiology 144:3257-3266.

Nicholas K.B., Nicholas H.B. and Deerfield D.W.1997. GeneDoc: analysis and visualization of genetic variation. EMB News, 4:14.

Nirvan R.S. 1960. Effect of antibiotic spray on citrus canker. Horticultural Advance (India) 4: 155-160.

Olson G.J. 1995. Eubacteria in the tree of life.

Padmanabhan D.P., Vidyasekran C.K. and Soumini Rajagopalan 1974. Changes in photosynthesis and carbohydrates content in canker and halo regions in X. citri infected citrus leaves. Indian Phytopath, 27: 215-218.

Padmanabhan S.Y. 1969. A mass screening technique against bacterial blight of rice caused by Xanthomonas oryzae. Indian Phytopath. 22: 396-398.

Pandey K.R. and V. Iswaran 1982. Influence of microbial fertilizers on bacterial blight of rice.II. Nitrogen content and extent of bacterial invasion of the rice leaves. Indian Phytopath. 35: 686-687.

Perego M. 1998. Kinase-phosphatase competition regulates Bacillus subtilis development. Trends Microbiol. 6:366-370.

Pettijohn D. E. 1990. Bacterial chromosome structure. Nucleic Acids Mol. Biol. 4:152-162.

Pettijohn D. E. and R. R. Sinden 1985. Structure of the isolated nucleoid, p. 199-227. In N. Nanninga (ed.), Molecular cytology of Escherichia coli. Academic Press, London, England.

Philip R. and S. Devadath 1980. Relation between some leaf characters and the occurrence of bacterial blight of rice. RISO 29: 317-322.

Raju B.C. and J.M. Wells 1986. Diseases caused by fastidious xylem-limited bacteria and strategies for management Plant Disease 70:182-86.

Rangarajan M. and B.P. Chakravarti 1970. Studies on the control of bacterial stalk rot of maize with antibiotics. Hindustan Antibiot Bull., 13: 16-19.

Rangaswami G., R.R. Rao and A. Lakshmanan 1959. Studies on the control of citrus canker with streptomycin. Phytopathology 49: 221-226.

Rönner S., Liesack W., Wolters J. and Stackebrandt E. 1991. Cloning and sequencing of a large fragment of the atpD-gene of Pirellula marina - a contribution to the phylogeny of Planctomycetales. Endocytobios Cell Res, 7:219-229.

Rosario M.M., Kirby J.R., Bochar D.A. and Ordal G.W. 1995. Chemotactic methylation and behaviour in Bacillus subtilis: role of two unique proteins, CheC and CheD. Biochemistry (Moscow), 34:3823-3831.

Ryder L., G. J. Sharples and R. G. Lloyd 1996. Recombination-dependent growth in exonuclease-depleted recBC sbcBC strains of Escherichia coli K-12. Genetics 143:1101-1114.

Sabet K.A. 1957. On the effect of certain environmental conditions on infection with the bacterial root and stalk rot disease of maize. Indian J. Agric. Sci. 27: 467-474.

Sanne Gowda. S. and G. Rangaswami (1963). "Studies on the virulence and pectolytic activity of the soft rot bacterium Erwinia carotovora (Jones) Holland", Indian Phytopath., 15: 226-231.

Saxena S.C. and S. Lal. 1986. Survival of Erwinia chrysanthemi pv. lege in soil. Indian J. Mycol. Plant Pathol. 16: 3741.

Schlesner H. and Stackebrandt E. 1986. Assignment of the genera Planctomyces and Pirella to a new family Planctomycetaceae fam. nov. and description of the order Planctomycetales ord. nov. Syst Appl Microbiol, 8:174.

Schmid M., Twachtmann U., Klein M., Strous M., Juretschko S., Jetten M., Metzger J.W., Schleifer K.H. and Wagner M. 2000. Molecular evidence for genus level diversity of bacteria capable of catalyzing anaerobic ammonium oxidation. Syst Appl Microbiol, 23:93-106.

Senecoff J.F. and Meagher R.B. 1993. Isolating the Arabidopsis thaliana genes for de novo purine synthesis by suppression of Escherichia coli mutants I. 5'-phosphoribosyl-5-aminoimida-zole synthetase. Plant Physiol, 102:387-399.

Setlow P. 1995. Mechanisms for the prevention of damage to DNA in spores of Bacillus species. Annu. Rev. Microbiol. 49:29-54.

Setlow B. and P. Setlow 1996. Role of DNA repair in Bacillus subtilis spore resistance. J. Bacteriol.178:3486-3495.

Shanna S.C., P.S. Randhawa, B.S. Thind and A.S. Khera 1982. Use of Klorocin for the control of bacterial stalk rot and its absorption, translocation and persistence in maize tissue. Indian J.Mycol. Plant Pathol. 12: 185-189.

Shekhawat G.S. and D.N. Srivastava 1972. Mode of action in bacterial leaf streak of rice and histology of the diseased leaf. Phytopath. Z., 74: 84-90.

Shekhawat G.S., D.N. Srivastava and Y.P. Rao 1972. Host specialization in bacterial leaf streak pathogen of rice. Indian J. Agric. Sci. 42: 11-15.

Singh R.S. 1989. Plant Diseases Sixth Edition, Oxford & IBH Publishing Co. Pvt. Ltd., New Delhi.

Singh D.V., A.K. BanerJee, R. Kishun and A.B. Abidi 1980. Effect of bacterial leaf streak on the quantitative and qualitative characters of (rice. Indian J. Mycol. Plant Pathol. 10: 67-68.

Singh R.N. 1972. Perpetuation of the bacterial blight disease of rice in north India. Indian Phytopath.25: 148-150.

Singh R.P., H.D. Chowdhury and J.P. Verma 1977. Races of Xanthomonas malvacearum. (Abs.),Indian Phytopath., 30(4): 572.

Singh Harnam and Nirmaljit Singh 1982. Efficacy of different chemicals in controlling common scab of potato. Indian J. Mycol. Plant Pathol. 12: 337-338.

Singh R.P., J.P. Venna and Y.P. Rao 1970. Eradication of seed infection of black arm of cotton. Curr. Sci. 39: 330-331.

Sitaramaiah K. and S.K. Sinha 1983. Relative efficacy of some selected antibiotics on bacterial wilt (Pseudomonas solanacearum biotype 3) of brinjal. Indian J. Mycol. Plant Pathol. 13: 277-281.

Sittig M. and Schlesner H. 1993. Chemotaxonomic investigation of various prosthecate and/or budding bacteria. Syst Appl Microbiol, 16:92-103.

Sivaswamy S.N. and A. Mahadevan 1986. Effect of stable bleaching powder on growth of Xanthomonas compestris pv. oryzae. Indian Phytopath. 39: 32-36.

Srinivasan N. and R.A. Singh 1982. Distribution pf Xanthomonas campestris pv. oryzae cells in rice tillers affected with Kresek phase of bacterial blight Indian Phytopath. 35: 147-148.

Srivastava D.N., Y.P. Rao, J.C. Durgapal and J.K. Jindal 1967. Bacterial leaf streak of rice in India. l. Dis. Reptr, 51: 928-929.

Srivastava D.N. 1969. Bacterial blight and leaf streak diseases of rice. Indian Phytopath. Soc. Bull.5: 64-66.

Starr M.P. 1959. Bacteria as plant pathogens. Ann. Rev. Microbiol. 13: 211-238.

Stragier P. and R. Losick 1996. Molecular genetics of sporulation in Bacillus subtilis. Annu.Rev. Genet. 30:297-341.

Strobel G .A. and W.N. Hess 1968. Biological activity of a phytotoxic glycopeptide produced by Corynebacterium sepedonicum. Pl. Physlol, 43: 1673-1688.

Strous M., Fuerst J.A., Kramer E.H.M., Logemann S., Muyzer G., van de pas-Schoonen K.T., Webb R., Kuenen J.G. and Jetten M.S.M. 1999. Missing lithotroph identified as new planctomycete. Nature, 400:446-449.

Suryanarayana D. and M.C. Mukhopadhyaya 1971. Ear cockle and 'Tundu', diseases of wheat, Ind. J.Agr. Sci., 41: 407-413.

Taylor P.A. and S.P. Flet 1981. Effect of irrigation on powdery scab of potatoes. Australian Plant Pathol. 10: 5556.

Teplyakov A., Obmolova G., Badet-Denisot M.A. and Badet B .1999. The mechanism of sugar phosphate isomerization by glucosamine-6-phosphate synthase. Prot Sci, 8:596-602

Thind B.S. and M.M. Payak 1985. A review of bacterial stalk rot of maize in India. Trop. Pest Management 51: 311.316.

Thind B.S. and P.S. Soni 1983. Persistence of chlorine in maize plant and soil in relation to control of bacterial stalk rot of maize. Indian Phytopath. 36: 687-690.

Thirumalachar M.J. 1968. Antibiotics in the control of plant pathogens. Advan. .Appll. Microbiol., 10: 313.

Traxler B.A. and Minkley E.G. 1988. Evidence that DNA helicase I and oriT site-specific nicking are both functions of the F Tra I protein. J Mol Biol, 204:205209.

Tri Murthy V.S. and S.Devadath 1982. Survival of Xanthomonas campestris (pv.. oryzy in different soils. Indian Phytopath. 35: 32-38.

Trun N. J. and J. F. Marko 1998. Architecture of a bacterial chromosome. ASM News 64:276-283.

Tvermyr M., Kristiansen B.E. and Kristensen T.1998. Cloning, sequence analysis and expression in E. coli of the DNA polymerase I gene from Chloroflexus aurantiacus, a green nonsulfur eubacterium. Genet Anal. 14:75-83.

Ullasa B.A., Y.R. Mehta, M.M. Payak and B.L. Renfro 1970. Xanthomonas rubrilineanson Zea mays in India. Indian Phytopath. 20: 77-78.

Vanoni M.A. and Curti B. 1999. Glutamate synthase: a complex iron-sulfur flavoprotein. Cell Mol Life Sci, 55:617-638.

Vasudeva R.S. and M.K. Hingorani 1952. Tannan or Tundu disease of wheat. Indian Farming. 2: 14-19.

Vasudeva R.S. and M.K. Hingorani 1952. Bacterial disease of wheat caused by Corynebacterum tritici. Phytopathology 42: 291-293.

Vergin K.L., Urbach E., Stein J.L., DeLong E.F., Lanoil B.D. and Giovannoni S.J. 1998. Screening of a fosmid library of marine environmental genomic DNA fragments reveals four clones related to members of the order Planctomycetales. Appl Environ Microbiol, 64:3075-3078.

Verma J.P. and R.P. Singh 1970. Two new races of Xanthomonas malvacearum, the incitant of black arm of cotton. Cott. Gr. Rev. 47: 203-205.

Verma J.P., M .L. Nayak and R.P. Singh 1977. Elimination of Xanthomonas malvacearum from cotton seeds by hot water. Ind. Phytopath., 30: 73-77.

Ward N.L., Rainey F.A., Hedlund B.P., Staley J.T., Ludwig W. and Stackebrandt E. 2000. Comparative phylogenetic analyses of members of the order Planctomycetales and the division Verrucomicrobia: 23S rRNA gene sequence analysis supports the 16S rRNA gene sequence-derived phylogeny. Int J Syst Evol Microbiol, 50:1965-1972.

Ward-Rainey N., Rainey F.A., Wellington E.M.H. and Stackebrandt E.1996. Physical map of the genome of Planctomyces limnophilus, a representative of the phylogenetically distinct planctomycete lineage. J Bacteriol 178:1908-1913.

Weinhold A.R., J.W. Oswald, T. Bowman, J. Bishop and D. Wright 1964. Influence of green manures and crop rotation on common scab of potato. Amer. Potato J. 41: 265-273.

Woese C.R., Olsen G.J., Ibba M. and Söll D. 2000. Aminoacyl-tRNA synthetases, the genetic code, and the evolutionary process. Microbiol Mol Biol Rev 64:202-236.

Woese C.R., Stackebrandt E., Macke T.J. and Fox G.E. 1985. A phylogenetic definition of the major eubacterial taxa. Syst Appl Microbiol, 6:143-151.

Woese D.J. 1987. Towards a natural system of organisms: Proposal for the domains archaea, bacteria, and eucarya. Microbiological Reviews 51:221-227.

Woldringh C. L., and N. Nanninga 1985. Structure of nucleoid and cytoplasm in the intact cell, p. 161-197. In N. Nanninga (ed.), Molecular cytology of Escherichia coli. Academic Press, London, England.

Zinder N.D. and J. Lederberg 1952. Genetic exchange in Salmonella, J. Bacteriol., 64: 679.

Index

A

Acetobacter xylinum, 16
Achizosaccharomyces, 78
Actionomycetes, 30
Activities of bacteria
  antibiotic resistance, 124
  bacterial strains, 127
  – transformation, 122
  coats, 98
  coinheritance frequency, 110
  complex transposons, 142
  bacteriophage mu, 144
  deletion generation, 143
  element deletion, 143
  polarity, 143
  regulation of transposition, 143
  transposition, 142
  conjugation, 112, 113, 118
  conservative transposition, 139
  cortex, 98
  enzymology, 111
  F lives two lives, 117
  function, 93
  genetic recombination, 103, 108
    homologous recombination, 108
    site-specific recombination, 108
  gliding movement, 94
  Gram reaction, 79-82
  host mutation descriptions, 125
deoR, 125
Dlon, 126
endA, 126
F′ F′, 126
galK, 126
galT, 126
gyrA, 126
hfl, 126
lacY, 126
lacZ, 126
lacZbM15, 126
locI, 126
malA, 126
proAB, 126
rec, 126
recA, 126
recJ, 126
relA, 126
rspL, 126
sbcBC,127
strA, 127
supE, 127
supF, 127
thi-1, 127
traD36, 127
umuC, 127
uvrC, 127
xylA, 127

integration of F, Episome, 118
intermolecular recombination, 106
modes of motion, 82-92
probability for recombination, 110
recombination frequency, 109
restriction/modification bacterial strains, 129
S cells, 122
spirochaetal movement, 93-94
spore formation, 95-96
sporulation, 99, 100
stages of spore formation in bacteria, 101
structure, 92-93
sunglling, 118
transduction, 134
transformation, 112
transposition, 139
transposon, 140, 141
chromosomal deletions, 142
polarity, 142
precise deletion, 142
regulation of transposition, 141
transposition, 141
transduction, 112
twitching movement, 95
Aerobic lithotrophs, 41
– organotrophs, 41
Aggregate plasmids, 67
Agrobacterium tumefaciens, 148
Agrobacterium, 88, 147
Agrostis alba, 172
Amphitrichous, 4
Anaerobic dissimulatory organotrophic, 46
– organotrophs, 47
– phototrophs, 47
Aquifex, 56
Atmospheric and temperature requirements of bacteria, 11
Atrichous, 4
Azotobacter, 102

B

Bacillus, 96, 133
Bacillus anthracis, 16
B. cereus, 99
B. subtilis, 77, 99, 136, 137
Bacteria
Chemoautotrophic, 7
Chemoheterotrophic, 7
encapsulated, 16
Gram positive and Gram negative, 7
Mesophilic, 9
Photochemosynthetic, 6
Psychrophilic, 9
rod-like, 2
spherical, 2
spiral-shaped, 2
structure of, 17-23
Thermophilic, 9
Bacterial Abundance
high diversity, 1
high total numbers, 1
shape
Bacterial capsules, 16
Bacterial diseases of plants, 145
affected foliage plants, 167
– ornamental plant, 167
agrobacterium species, 185
crowngall of stone fruits, 185
bacterial canker in *citrus* leaves, 156, 157
– pathogenes, 151
clavibacter species: 186
bacterial canker in tomato, 190
– ear rot of wheat, 188
common control measures, 149
– symptoms of bacterial plant diseases, 147

local lesions, 147
scabs and cankers, 148
soft rots, 147
vascular disease, 147
tumours and galls, 148
curtobacterium species, 191
bacterial wilt of beans, 192
erwinia species, 178
bacterial blight of foliage plants, 184
– stalk rot of maize, 178
– wilt of cucurbits, 182
black leg and soft rot of potato, 180
fire blight of apple and pears, 182
stem rot of ornamental plants, 184
pseudomonas species, 170
bacterial brown rot of potatoes, 171
– leaf spot of foliage plants, 178
– speck of tomato, 170
brown spot of beans, 176
halo blight of beans, 174
leaf spot of dracaena sanderana, 177
– – – ornamental plants, 177
pseudomonas rubrilineans, 173
red stripe of sugarcane, 173
wild fire of tobacco, 173
wilt disease of potatoes, 171
Ralstonia solanacearum, 195
rhodococcus species, 192
abnormal branching of ornamental plants, 193
streptomyces species, 193
common scab of potato, 193
xanthomonas species, 150
angular leaf spot, 150
bacteria blight of cotton, 150
– – – of foliage plants, 169
– – – beans, 164
– – – ornamental plants, 168
– – – streak of rice, 160
– – – spot of foliage plants, 167
– – – tomato, 162
– leaf blight of rice, 158
black arm of cotton, 150
black rot of crucifrs, 163
citrus canker, 156
leaf spot of mango, 164
Bacterial evolution, 3
– plasmids, 8
– polysaccharides, 15
– reproduction and genetic recombination, 8
– ribosome, 24
Bergey's manual of determinative bacteriology, 27
Bi-directional replicative transposition, 140
Burrill, T.J., 2

C

Capsules, 15
macrocapsules, 15
microcapsules, 15
Classification of bacteria, 27
Actinomycetes with multilocular sporangia, 49
Actinoplanetes, 49
Bacterial nomenclature, 56
Bacteroides/flavobacteria, 53
Cell wall-less, 48
characteristics, 27
chemical composition of the cell, 28
chlamydias, 52

chloroplasts, 51
cyanobacteria, 51
cytophaga, 53
flexibacter, 53
Gram-positive bacteria, 46, 48
high GC, 48
kingdom-procaryoteae, 28
maduromycetes, 50
nature of DNA, 27
nineteen sections of bacterial group by bergey's manual, 28
photobacteria, 28
red photobacteria, 28
green photobacteria, 28
mollicutes, 28
scotobacteria, 28
rickettsias, 28
nocardioforms, 49
other cytological features, 28
planctomyces/pirella, 52
Prochlorophytes and chloroplasts, 52
prochlorophytes, 51
sporocytophaga, 53
sterptomyces and related genera, 50
thermoactinomycetes, 51
thermonospora and related genera, 50
verrucomicrobia, 53
Clavibacter xyli, 187
Clostridium, 96
Composite plasmids, 66
Conjugant cells, 116
Corynebacterium flaccumfaciens, 192
C. *insidiosum*, 187
C. *iranicum*, 187
C. *michiganese*, 186
C. *nebraskense*, 187
C. *poinsettiae*, 192
C. *rathayi*, 187
C. *renale*, 90
C. *sepedonicum*, 187
C. *tritici*, 187
C. *renale*, 88
Curtobacterium, 147
Cyanobacteria, 102
Cytoplasm, 13

D

D- and L-amino acids, 20
Deferribacter group, 56
Deinococci, 54
Desulfotomaculum, 96
D-galactose, 16
D-glucose, 16
D-glucuronic, 16
D-mannose, 16
DNA, 14, 15, 26, 58, 59, 60, 62, 64, 67, 68, 69
DNA binding, 97
DNA characteristics of bacteria, 58
Control of gene expression in bacteria, 71
regulating enzyme activity, 71
– – synthesis, 71
operons and their control, 71-72
F-factor or sex factor, 69
gene discovery in domain bacteria, 73-78
inducible operon, 72-73
large scale isolation of plasmid and bacterial DNA, 61
nucleoid, 59
repressible operon, 72
R-plasmids or R-factor, 67
Donar cell, 116
DsDNA, 107

E

Escherichia coli (E. coil), 7, 24, 89, 103, 113, 115, 123, 124, 125, 135, 136, 138

Endospore forming rods and cocci, 30

– structure, 96

Epipremnum aureum, 178

Erwinia amylovora, 182

F

F^+ cell, 117

Facultatively aerobic, 47

Fimbriae, 88-91

Flagella, 4, 82-87

Fluid-mosaic, 18, 19

G

Gemmata obscuriglobus, 73, 74, 75, 76, 77, 78

Gene homologs, 73, 76

Glutamine oxoglutarate aminotransferase (GOGAT), 76

Glycocalyx, 92

Gram, Hans, 7

Gram negative, 8, 18, 41, 46, 80

Gram positive, 8, 18, 30, 80

Gram stain of Gram, 7

Gram-positive, 186

Green nonsulfur bacteria, 55

– sulfur bacteria, 54

Groups of bacteria, 40

acetic acid bacteria, 42

azotobacter, 42

budding and prosthecate/stalked bacteria, 45

enteric bacteria, 43

gliding myxoacteria, 45

hydrogen-oxidizing bacteria, 41

methanotrophs and methyltrophs - methlyosinus, methylcoccus, 41

neisseria, chromobacterium and Relatives, 43

nitrifying bacteria, 40

proteobacteria, 40

pseudomonas and the Pseudomonads, 41

purple phototrophic bacteria, 40

rickettsias, 44

sheathed proteobacteria, 44

spirilla, 44

sulfate and Sulfur-reducing bacteria, 45

sulfur and iron-oxidizing bacteria, 40

vibrio and photobacterium, 43

H

Haemophilus, 133

Heteropolysaccharide, 16

H-ion concentration, 8

I

Impact of External Environment on Bacteria, 8

light, 10

mesophilic Bacteria, 9

moisture, 10

pressure, 10

psychrophilic Bacteria, 9

temperature, 8-9

thermophilic Bacteria, 9

Indicator plasmids, 65

International code on bacterial nomenclature, 57

International Committee on Systematic Bacteriology (ICSB), 57

Interrupted-making, 71

Intracellular organelles, 15

L

Lake's model, 24

L-amino, 16

Lederberg, Joshua, 117

Leeuewenhock, Antony Von, 1
Leuconostoc mesenteroides, 16
Lipid-lipid, 18
Lipid-protein, 18
Lophotrichous, 4
L-rhamnose, 16
Lux operon, 132
Lysogenes, 59

M

M13 viral transfection, 124
Mechanism of conjugation in *E.coli*, 114
Methane-producing bacteria, 30
Microaerophilic, 47
Monotrichous, 4
Mycobacterium leprae, 81
M. *tuberculosis*, 81, 103
Mycoplasmas, 30
Myxococcus, 102

N

N-acetyl glucosamine, 20
N-acetyl muramic acid, 20
Neisseria gonorrheae, 91
Nitrogen fixing bacteria, 145
Nitrospira group, 56
Nobel prize, 117
Non-self-transmissible plasmids, 66
Nutrition
 chemoautotrophs, 6
 classification, 6
 methods of, 6
 mode of 5
 parasites, 6
 photoautotrophus, 6
 saprobes, 6

O

Outer membrane proteins (OMPs), 81

P

Pasteur, 1
Peritrichous, 4
Philodendron panduraeforme, 178
Photosynthetic bacteria, 6
Physical forces, 8
Pirellula marina, 73, 74, 75, 76, 77, 78
Planctomyces limnophilus, 75
Plasmids, 63
Polyglucan granules, 23
Poly-b-hydroxy butyrate granules, 23
Porins, 81
Pseudomonas aeruginosa, 16
P. cichorri, 177
P. syringae, 174, 176

Q

Quellung reaction, 16

R

R factors, 91
Recipient cell, 116
Relationship between
 bacterial growth and oxygen, 11
 bacterial growth and temperature, 12
Relatives, 56
Respiration, 5
Rhizobium iupini, 90
Rhodococcus, 147
Rhodospirillum, 93
Ribosomes, 23
Rickettsias, 30
RNA, 14, 60, 104

S

Salmonella typhimurium, 88
Self-transmissible plasmids, 66
Shigella flexnerii, 88

Singer-Nicolson membrance, 19
Solanum tuberosum, 172
Spirillum volutans, 23
Spirochetes, 54
Sporocarcina, 96
Stofller and wittmann's model, 24
Streptococcus pneumoniae, 121, 133
S. *typhimurium*, 136
S. scabies, 193
Streptomyces, 147
Strictly anaerobic anoxygenic phototrophs, 40
- purple nonsulfur bacteria, 40
- purple sulfur bacteria, 40

Stringent and relaxed plasmids, 66
Structure of bacteria, 13
- bacterial cell wall, 20
  - carbohydrate components, 20
  - peptidoglycans, 22
  - protein components, 21
- bacterial cytoplasm, 23
- cytoplasmic member, 17
  - chemotaxis, 18
  - electron transport and oxydative phosphorylation, 17
  - genetical approach, 18
  - photosynthesis, 18
  - synthesis of cell wall components, 17
  - transport of nutrients, 17
- Mesosomes, 25

T

Thermal death, 9
Thermocrinus, 56
Thermodesulfobacterium, 55
Thermotoga, 55
Transformation of bacteriophage, 124
TRNA, 25

V

Vegetative cells, 9

X

Xanthomonas campestris, 166, 167, 195
X. *erwinia*, 147